# STAR
# WATCH

The Amateur Astronomer's
Guide to Finding, Observing,
and Learning about
Over 125 Celestial Objects

## Philip S. Harrington

WILEY

John Wiley & Sons, Inc.

Published by John Wiley & Sons, Inc., Hoboken, New Jersey
Published simultaneously in Canada

For general information about our other products and services, please contact our Customer Care Department within the United States at (800) 762-2974, outside the United States at (317) 572-3993 or fax (317) 572-4002.

Wiley also publishes its books in a variety of electronic formats. Some content that appears in print may not be available in electronic books. For more information about Wiley products, visit our web site at www.wiley.com.

*Library of Congress Cataloging-in-Publication Data:*

Harrington, Philip S.
    Star watch : the amateur astronomer's guide to finding, observing,
  and learning about over 125 celestial objects / Philip S. Harrington.
        p. cm.
    Includes bibliographical references and index.
    ISBN 978-1-63026-169-6
    1. Astronomy—Amateurs' manuals. 2. Astronomy—Observers' manuals.
  I. Title.

  QB63.H319 2003
  522—dc21

                                                                2003043272

10  9  8  7  6  5

*For my niece Lilia, nephew Daniel,*
*and goddaughter Samantha.*
*In loving memory of Dorothy Harrington (1919–2003).*

# Contents

# Preface

*Astronomy . . . is most impressive where it transcends explanation. It is not the mathematics of astronomy, but the wonder and mystery that seize upon the imagination. The dominion which astronomy has always held over the minds of men is akin to poetry; when the former becomes merely instructive and the latter purely didactic, both lose their power over the imagination. Astronomy is known as the oldest of the sciences, and it will be the longest-lived because it will always have arcana that have not been penetrated.*

With those words, Garrett P. Serviss opened his book *Curiosities of the Sky.* Though published in 1909, those words remain true to this day. The wonder of astronomy is not in the numbers and data that scientists collect and study, but rather in what is behind the numbers: the amazement and curiosity that the objects of the universe generate in our minds. Serviss's book was meant to take the reader to those wonders, to explain what each was, and to demonstrate what the backyard astronomer of the early twentieth century might see through a telescope. To that end, Serviss continued: "Some of the things described in this book are little known to the average reader, while others are well known; but all possess the fascination of whatever is strange, marvelous, obscure, or mysterious—magnified, in this case, by the portentous scale of the phenomena." I cannot think of a better way to describe the purpose of this book that you hold before you. We will follow the same path trod by Serviss one hundred years ago, but through the eyes of a twenty-first century astronomer using modern equipment. Wondrous sights, including the Moon, Sun, and planets, star clusters, nebulae, and far-distant galaxies are up there waiting for us, as they have been for millions, even billions, of years.

But the universe is a far different place from what Serviss could have ever imagined. Today, thanks to space probes to the Moon and planets, huge telescopes here on Earth, orbiting astronomical observatories, and advances in optics and electronics, astronomers have uncovered some amazing facts about our universe. As amateur astronomers, we can enjoy many of these same advances, both in knowledge and equipment, allowing us to view sights that Serviss never knew existed.

This book will serve as a guide for you to explore these far-distant objects. Chapter 1 explains some basic astronomical terms that will help orient the newcomer to the sky. Beginning with chapter 2, we launch our trip by taking one small step out to the Moon. In the chapters that follow, our attention will be turned toward the planets, the Sun, and finally the four seasonal skies, each of which holds some outstanding deep-sky objects—objects that lie beyond our solar system—including vast star clusters, glowing nebulae, and distant galaxies. Each target, whether near or far, is discussed both in terms of locating and observing it, as well as in terms of the latest scientific studies and findings. *Star Watch* gives you over 125 suggested targets for viewing and tells you how to find them, when to view them at their best, and what you can expect to see through both binoculars and telescopes. You'll also find physical descriptions of each object, including size, distance, and structure, as well as scientific explanations of what you are looking at.

Remember, large telescopes are *not* needed to discover the heavens. Indeed, all of the objects discussed in this book can be spotted through a modest 6-inch telescope; in fact, all but perhaps half a dozen can be viewed through instruments half that size and even through common binoculars. Garrett Serviss would have been amazed!

I welcome your comments about this book, whether they be praise or complaints. Just write to me in care of the publisher, John Wiley & Sons, or directly via e-mail at phil@philharrington.net.

# Acknowledgments

Several people have been instrumental in creating *Star Watch,* and it is only proper to acknowledge their contributions and express my gratitude to each in this forum. All the celestial photographs found throughout this book were taken by amateur astronomers. Astrophotography is not easy, and so I must thank those accomplished photographers who graciously allowed me to use some of their work: Jim Fakatselis, Chris Flynn, Brian Kennedy, Richard Sanderson, Gregory Terrance, and George Viscome.

I wish to pass on my sincere appreciation to my astronomical colleagues who read through the manuscript and offered many constructive suggestions: Chris Adamson, Kevin Dixon, Geoff Gaherty, and Richard Sanderson. It is difficult enough to critique a stranger's written word, but when a friend asks you, it can be especially touchy. Just how honest do you want to be? Let me say that each of these readers was both sensitive to my fragile ego, as well as aware that changes were needed to make a better book. You and I are both in debt to them for their ability to balance both needs. Special thanks to Kate Bradford, my editor at John Wiley & Sons, for suggesting the idea for this book in the first place, and for her diligent guidance and help.

Finally, my deepest thanks, love, and appreciation go to my ever-patient family. My wife, Wendy, and our daughter, Helen, have continually provided me with boundless love, patience, and encouragement over the years. Were it not for their understanding my need to go out at three in the morning or drive an hour or more from home just to look at the stars, this book could not exist. I love them both dearly for that.

# 1

# Your Passport
# to the Stars

Since before the dawn of history, our ancestors have gazed skyward in awe and wonder. At first, the universe seemed cold, inhospitable, and filled with danger. Was the sky populated by gods of the Moon, the Sun, and planets who ruled mortals and could destroy or spare Earth with a mere wave of their hands? Did our very lives depend on their whimsy? As time evolved, curiosity grew about the exact nature of the Sun, the Moon, and other sky objects. The ancient Babylonians, Assyrians, and Greeks were among the first civilizations to study the sky in an attempt to understand how these objects influenced the course of human events. Their studies gave birth to the pseudoscience of astrology but more importantly also laid the basic foundation for the science of astronomy. Indeed, many of the names for stars and star patterns, called *constellations,* that we still use today trace their origin back to this early epoch.

Perhaps the greatest study of the universe by one of our ancient ancestors was performed by the Greek astronomer Claudius Ptolemaeus, or Ptolemy for short. While living in Egypt in the second century A.D., Ptolemy devised a scheme that predicted the movements of the Sun, the Moon, and planets in our sky with amazing accuracy. Ptolemy's *geocentric* system, which placed Earth at the center of the universe, remained the dominant model for more than a thousand years, until the European Renaissance in the sixteenth and seventeenth centuries.

Today, we know the true order of the universe. No longer populated by fearsome gods and goddesses, our universe plays host to stars and galaxies, planets and moons, and many other wonders that beckon us to stare skyward with the same awe and wonder felt by the first astronomers. We know that Earth is not at the center of the universe, and in fact that the universe really has no center at all. Instead, the universe is populated by millions, if not billions, of *galaxies* (Figure 1.1), each a huge system of stars. With rare exception, all galaxies are seemingly racing away from one another, motions induced by the *Big Bang*, which created the universe some 15 billion years ago. Some are independently traveling through the universe, while others travel in groups and clusters.

Each galaxy, including our own Milky Way, is made up of millions, if not billions, of individual stars. Many stars exist alone, while others formed in pairs, trios, or larger groupings. The largest star groupings are referred to as *star clusters* (Figure 1.2). Depending on the type, a cluster may hold anywhere from a dozen to half a million stars!

Interspersed throughout many galaxies are large clouds of gas and dust called *nebulae* (Figure 1.3). Some nebulae may be thought of as stellar nurseries, marking regions where new stars are forming. Other nebulae are stellar corpses, all that remain of once powerful suns.

Closer to home, our star, the Sun, serves as the focal point of a collection of comparatively small celestial bodies that we collectively call the *solar system*. In addition to the Sun, the solar system includes the nine planets—Mercury, Venus, Earth, Mars, Jupiter, Saturn, Uranus, Neptune, and Pluto—as well as many moons, asteroids, comets, and meteoroids.

**Figure 1.1** *Spiral galaxy M81 in the constellation Ursa Major. This photograph, taken by George Viscome through a 14.5-inch f/6 reflector, is oriented with south toward the top to match the view through most astronomical telescopes.*

a

b

**Figure 1.2** *(a) Open star cluster M36 in the winter constellation Auriga. (b) Globular cluster M22 in Sagittarius. South is toward the top of each of these photographs taken by George Viscome through a 14.5-inch f/6 reflector.*

a

**Figure 1.3** *(a) Bright nebula M16, better known as the Eagle Nebula, in the summer constellation Serpens. (b) Planetary nebula M76 in the autumn constellation Perseus. South is up in each of these photographs taken by George Viscome through a 14.5-inch f/6 reflector.*

b

4

All of these objects are waiting for you in the sky. But gazing skyward, trying to identify one from another or simply knowing where to look, can prove to be a daunting task for someone brand-new to the universe! Where is everything? Which star is which? Are there any planets out tonight? I just bought a new telescope; what should I look at first? And how do I use it?

All good questions. Where do you find the answers? I hope you find them here! Think of this book as your passport to places that few of the world's population are even aware exist. You and I are heading off to the farthest depths of space.

## The Sky: An Overview

At first glance, the sky appears to be a mishmash of points of light, seemingly scattered at random and impossible to fathom. But the sky isn't as chaotic as it seems at first. It has a clear order that just takes a little time to understand. Let's begin with some basics.

### Sky Motions

Watching the stars move silently overhead at night, we can easily see how Ptolemy and other ancient skywatchers came away with the idea that everything circled Earth. Instead, we know that this effect is caused by Earth rotating on its *axis*, an imaginary line that passes through the center of Earth, with the North Pole at one end and the South Pole at the other. This motion, called *daily* or *diurnal motion* (Figure 1.4), causes the Sun, the Moon, planets, and stars to appear as though they rise in the east, move across the sky, and set in the west. Daily motion also opens our sky window toward different stars at different times of the night.

The only stars that do not appear to rise and set are those located near the two *celestial poles,* the projection of Earth's North and South Poles against the sky. Instead, these *circumpolar* stars and constellations rotate above our horizon all night long. Which constellations are circumpolar from your location depends on your *latitude* (the angular distance you are away from the equator, measured in degrees). For most of us who view the sky from midnorthern latitudes (which includes most of the United States, southern Canada, and most of Europe), there are six generally recognized circumpolar constellations: Ursa Major, Ursa Minor, Cassiopeia, Cepheus, Draco, and Camelopardalis. If you live in Hawaii, southern California, and Texas, along the Gulf of Mexico, or in Florida, portions of some of these constellations may dip below the horizon at times, technically disqualifying them as circumpolar. If you live in Alaska, northern Canada, Europe, or Asia, then more constellations will be in your circumpolar zone. Indeed, from the North Pole, all constellations that are visible are circumpolar, since none rise or set (likewise from the South Pole, although the other half of the sky is visible).

In addition to rotating on its axis, Earth also revolves around the Sun, taking one year to complete the trip. We call this *annual motion*. It takes Earth 365 days 5 hours 49 minutes to complete a revolution of the Sun, at an

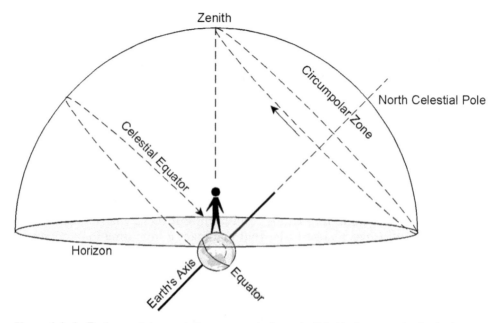

**Figure 1.4** *As Earth completes a rotation once each day, celestial objects appear to rise in the eastern part of the sky and set toward the west. Only stars that lie exactly on the celestial equator rise exactly in the east and set exactly west. Stars that are within the circumpolar region appear neither to rise nor set, but instead to remain above the horizon all night along. Centered in the circumpolar zone is the celestial pole, the projection of Earth's axis against the celestial sphere. Right now, the North Celestial Pole is aimed almost exactly at Polaris, the North Star.*

average speed of 66,500 miles (107,000 kilometers) per hour or 18.46 miles (29.73 kilometers) each second!

Annual motion also opens our sky window onto different stars and constellations at different seasons of the year. Those stars and constellations seen on winter evenings, for instance, are called the winter stars, and are completely different from those visible on summer evenings. During the summer and winter months, the night side of Earth is aimed toward different portions of the plane of our galaxy, where the gentle rifts of the Milky Way stretch across the sky to give spectacular views of rich star clouds and subtle nebulae. The spring and fall skies open away from the obscuring dust clouds of our galaxy to reveal a universe that is full of other distant galaxies. Figure 1.5 shows Earth in its annual trek around the Sun as compared with the seasonal constellations.

Finally, Earth also wobbles like a toy top in a motion called *precession.* If you've spun a top, you know that its axis of rotation is usually perpendicular to the surface on which it is spinning. But give it a slight nudge and the top will begin to tip over. Still spinning, the top attempts to right itself, in the process causing its axis of rotation to trace out a circle. The combined effect of Earth slightly bulging at the equator, combined with gravitational nudges from the Sun and the Moon, has nudged Earth enough to set it into a slow

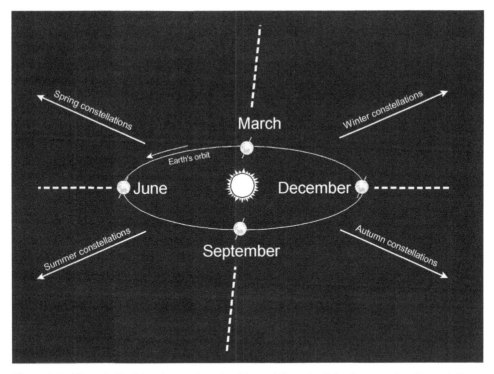

**Figure 1.5** *Although Earth's axis remains aimed toward the celestial poles, our planet's revolution around the Sun turns our night sky toward different parts of the universe during different times of the year. This diagram shows Earth's positions relative to the Sun and background constellations at four separate points during its orbit.*

wobble. Precession is barely noticeable compared to our planet's rotation and revolution, taking some 26,000 years for Earth to complete just one twist. While this shift cannot be detected with the naked eye over the brief course of a human lifetime, the change is obvious across the span of human history. Right now, the Earth's North Celestial Pole is aimed almost directly at the star Polaris, also known as the North Star. But back in the time of ancient Egypt, the pharaohs saw the night sky turning about the star Thuban in our constellation Draco the Dragon. In 13,000 years, the star Vega in the constellation Lyra will be closest to the North Celestial Pole.

### Star Brightness

When astronomers talk about how bright something appears in the sky, they are referring to that object's *magnitude*. The magnitude system dates back over 2,000 years to the Greek astronomer Hipparchus, who was first to survey the night sky and devise a system for categorizing stars according to their brightness. His method was quite simple. The brightest stars visible to the eye were labeled 1st magnitude, while the faintest stars were 6th magnitude. The remaining stars fell somewhere in between. In the Hipparchus magnitude system, the larger the magnitude number, the fainter the star. The magnitude system in place today is far more precise but still strongly reminiscent of Hipparchus's. We still use the basic 1st-through-6th magnitude designations, but we now specify that a 1-magnitude jump (say from 1st to 2nd, or 2nd to 3rd) corresponds to a change in brightness of 2.5 times. Therefore a 1st-magnitude star is 2.5 times brighter than a 2nd-magnitude star, while a 2nd-magnitude star is 2.512 times brighter than a 3rd-magnitude star. By this method, a 1st-magnitude star is about 6.3 times brighter than a 3rd-magnitude star (2.512 × 2.512 = 6.310), about 15.8 times brighter than a 4th-magnitude star (2.512 × 2.512 × 2.512 = 15.85), about 40 times brighter than a 5th-magnitude star (2.512 × 2.512 × 2.512 × 2.512 = 39.81), and so on. A 5-magnitude jump, say from 1st to 6th magnitude, equals a change in brightness of exactly 100 times.

Astronomers of the nineteenth century refined and expanded Hipparchus's magnitude system to include the very brightest and very faintest celestial objects, so the scale doesn't stop at 1st or 6th magnitude. A zero-magnitude object is 2.512 times brighter than a 1st-magnitude object, while negative-value objects are brighter still. The Sun, for instance, is magnitude −26. At the same time, stars that are too faint to be seen with the naked eye have magnitude values greater than 6th magnitude. As you can see from Figure 1.6,

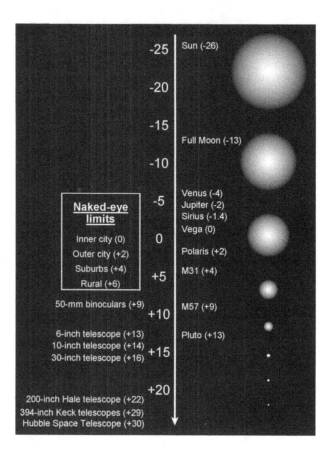

**Figure 1.6** *The magnitude scale. The Sun, the brightest object in the sky, rates magnitude −26, while the faintest stars ever photographed through the Hubble Space Telescope are rated magnitude +30. Out in the countryside, the human eye can detect stars as faint as magnitude +6 (possibly fainter under extraordinary conditions) but may only see stars to magnitude 0 or +1 from cities because of light pollution.*

binoculars reveal stars to about 9th magnitude, a 6-inch (15-cm) telescope to 13th magnitude, and larger telescopes deeper still. The Hubble Space Telescope can record stars as faint as 30th magnitude, about 19 billion times fainter than the Sun!

Magnitudes only refer to how bright objects appear in our sky. Just because a star looks bright in the sky doesn't mean that it is big, of course. For example, the Sun is the brightest object in our sky, but pull back an appreciable distance and it quickly disappears into the crowd. The Sun is only bright because it is so close. A star's apparent brightness, or magnitude, depends on two factors: its intrinsic luminosity and distance. To express a star's true brightness, astronomers use the term *luminosity*. The luminosity scale uses the Sun as its baseline value of 1.0. Stars with luminosities greater than 1.0 are intrinsically brighter than the Sun, while those less than 1.0 are dimmer. Sirius, the brightest star in the sky, has a luminosity of 24, meaning that it is 24 times more luminous than the Sun, while the star Deneb in Cygnus has a luminosity of 24,000. So why does Sirius look brighter in our sky than Deneb? Measurements show that Sirius lies only 9 light-years away, while Deneb is more than 3,200 light-years from us.

### Star Sizes and Distances

The sizes of stars, planets, and other objects are referred to in two different ways. One is a measure of their actual size. The Moon, for instance, measures 2,159 miles across (3,476 kilometers), while the Sun is 864,400 miles (1,392,000 kilometers) in diameter, and so on.

*Apparent size,* or how large something appears in the sky, is expressed in angular degrees, which can be further divided into arc-minutes and arc-seconds. Figure 1.7 offers an example. The Andromeda Galaxy apparently spans 5° in length in photographs. This is considerably more than the Moon or Sun, each of which measures half a degree, or 30 arc-minutes (abbreviated 30'). Sky objects that are even smaller than 1 arc-minute are measured in arc-seconds. There are 60 arc-minutes (60') in 1 degree and 60 arc-seconds in 1 arc-minute.

Values in miles and kilometers quickly become too cumbersome to use when talking about the actual distances to the stars, so astronomers refer to distances in *light-years*. One light-year is equal to the distance that a beam of light would travel in space in one Earth year, more than 5.87 trillion miles (9.45 trillion kilometers), or 186,000 miles per second (300,000 kilometers per second)! The Andromeda Galaxy, usually regarded as the most distant object visible to the naked eye, lies about 2.9 million light-years away. That means the light we would see from it tonight, traveling at 186,000 miles every second, took 2.9 million years to get here! That translates to a distance of 17 billion miles (27.4 billion kilometers)!

Astronomers use very precise tools to measure apparent sizes of and distances between objects in the sky with great accuracy, but you and I were born with a handy tool that we can use to approximate those same values. It's your hand! Take a look at Figure 1.8. It turns out that the ratio of the size of the human hand to the length of the arm is proportional for everyone,

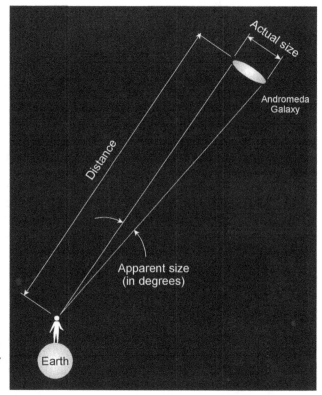

**Figure 1.7** *How large an object appears in our sky, called its apparent size, depends on two factors: its actual size and its distance away. Despite its great distance, the Andromeda Galaxy M31 spans close to 5° in our sky because it is physically very large.*

regardless of age or gender. For instance, at arm's length, your fist covers 10°. Your middle three fingers extended as in a Scout salute cover 5° of sky, the same span as the pointer stars at the end of the bowl of the Big Dipper. The span between your pinky finger and forefinger equals 15°, while the span between your thumb and pinky equals 25°. (Note that some people can stretch their hands more than others, which may throw this last measurement off a bit. To find out your hand span, hold it up against the Big Dipper. If your finger and thumb can cover its length fully, your span is 25°; a little less, and it's probably closer to 20°.)

When you get to the later chapters that discuss each seasonal sky, remember this "handy" method for finding distances between stars and constellations. It makes getting around the sky much easier.

### Star Positions

If you were to give me directions to your home, you might take me from a major highway or thoroughfare to a secondary road, and finally to your street and house number. That is pretty much how most amateur astronomers find objects in the sky. They begin at a major constellation, travel to a particular star in the pattern, then follow fainter stars to the target itself. This technique, called *star hopping,* is discussed later in this chapter.

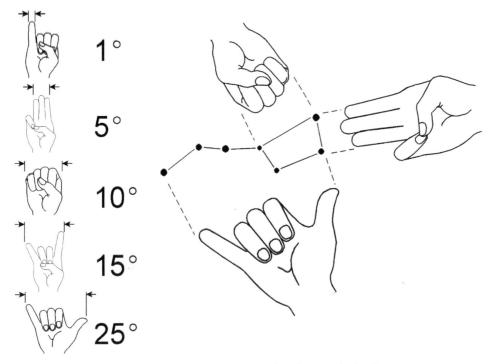

**Figure 1.8** *The human hand is a useful tool for estimating distances in the sky.*

While these methods work fine out in the field, they can be cumbersome when trying to list either an earthly location or celestial object in a data catalog. Instead, geographers have divided up Earth into a north-south and east-west coordinate system called latitude and longitude, respectively. Astronomers have similarly divided up the sky into a north-south, east-west coordinate system. Rather than use longitude and latitude, celestial coordinates refer to *right ascension* and *declination*.

Let's look at declination first. Just as latitude is the measure of angular distance north or south of the Earth's equator, declination (abbreviated Dec.) specifies the angular distance north or south of the *celestial equator*. The celestial equator is simply the projection of Earth's equator up into the sky. If we were positioned at 0° latitude on Earth (the equator), we would see 0° declination pass directly through the zenith, while 90° north declination (the North Celestial Pole) would be overhead from the Earth's North Pole. From our South Pole, 90° south declination (the South Celestial Pole) is at the zenith.

As with any angular measurement, the accuracy of a star's declination position may be increased by expressing it to within a small fraction of a degree using arc-minutes and arc-seconds:

$$1 \text{ degree } (1°) = 60 \text{ arc-minutes } (60')$$
$$= 3{,}600 \text{ arc-seconds } (3{,}600'')$$

Right ascension (abbreviated R.A.) is the sky's equivalent of longitude. The big difference is that while longitude is expressed in degrees, right ascension divides the sky into twenty-four equal east-west slices called "hours." Quite arbitrarily, astronomers chose as the beginning or zero-mark of right ascension the point in the sky where the Sun crosses the celestial equator on the first day of the Northern Hemisphere's spring. A line drawn from the North Celestial Pole through this point (the vernal equinox) to the South Celestial Pole represents 0 hours right ascension. Therefore, any star that falls exactly on that line has a right ascension coordinate of 0 hours. Values of right ascension increase toward the east by 1 hour for every 15° of sky crossed at the celestial equator.

To increase precision, each hour of right ascension may be subdivided into 60 minutes, and each minute into 60 seconds. In other words:

$$1 \text{ hour R.A. (1 h)} = 60 \text{ minutes R.A. (60 m)}$$
$$= 3{,}600 \text{ seconds R.A. (3,600 s)}$$

Unlike declination, where a minute of arc does not equal a minute of time, a minute of R.A. does.

A star's celestial coordinates do not remain fixed forever. Recall that Earth wobbles in a 26,000-year cycle called precession. Throughout the cycle, the entire sky floats behind the celestial coordinate grid. While this shifting is insignificant from one year to the next, astronomers find it necessary to update the stars' positions every fifty years or so. That is why you will notice that all right ascension and declination coordinates are referred to as "epoch 2000.0" in this book and most other contemporary volumes. These indicate their exact locations at the beginning of the year 2000, but they will remain accurate enough for most purposes for several decades to come.

## A Survey of the Sky

Many sky objects are perfect targets for binoculars and backyard telescopes. Some are better seen through telescopes, others through binoculars, but all have intrinsic beauty and interest.

When describing objects in the following chapters, I've attempted to answer five fundamental questions we all have when looking through a telescope:

1. What can I see?
2. When and where should I look?
3. How can I find it?
4. What does it look like?
5. What am I looking at?

### What Can I See?

The first question most new telescope owners ask is: "Okay, so what can I see through my telescope?" The short answer is "Plenty!" We begin by venturing into our immediate neighborhood, the solar system, with the first stop at

Earth's nearest neighbor in space, the Moon. Whether you use a telescope, binoculars, or gaze skyward with your eyes alone, the Moon is a familiar part of the night sky. It is the only object in the night sky that reveals surface details to the naked eye. Binoculars and telescopes expand on this to show a spectacular sight, with literally hundreds of lunar features coming into view as the Moon progresses through its phases. Witnessing for yourself the stark beauty of the Moon up close is an overwhelming experience.

Next, we will explore our solar system. Each of the five naked-eye planets—Mercury, Venus, Mars, Jupiter, and Saturn—has something to offer. Even if you only have a 2-inch (5-cm) telescope, you can see the phases of Venus, the moons and cloud belts of Jupiter, the rings of Saturn, the Martian polar caps, and even the distant disks of Uranus and Neptune. Only lonely Pluto is missed; for that, you will need at least a 6-inch (15-cm) telescope.

Our survey of the solar system would hardly be complete without a visit to our star, the Sun. But always be careful when viewing the Sun, since it is the only celestial object that can actually cause you physical harm. Without proper precautions, its intense radiation can burn your eyes' retinas just as it can burn your skin, but much more quickly. There are many ways to view the Sun safely, as described in chapter 4.

Finally, we leave the confines of our solar system to explore the seemingly limitless boundaries of the universe. Even the smallest binoculars and telescopes can show dozens of deep-sky objects, including binary and variable stars, star clusters, nebulae, and galaxies. And this book will get you to them!

### Star Identification

It's an awfully big universe, so before venturing out, let's look at how sky objects are identified. For thousands of years, stars have been grouped into large patterns that we call constellations. There are eighty-eight constellations in all, listed alphabetically in Appendix A. Our ancient ancestors created most of these fanciful star pictures as they gathered their families at night to tell exciting myths and legends about some amazing creatures and beings. Storytellers would illustrate their narratives using figures drawn among the stars. In reality, the stars that make up a constellation have no real relationship to one another and are more than likely tens or hundreds of light-years away from one another in space. Nobody knows who described the first constellation, but most come to us from ancient Greece, Rome, Egypt, Persia, and Babylonia. Other cultures, such as Native Americans and those of the Orient, also made up constellations, but they were quite different from the more familiar, western ones.

Astronomers use the traditional constellations as helpful guides for locating and naming stars and other sky objects. Every object has been assigned to a home constellation, even though it may not contribute to the constellation's figure. Picture the night sky as a community divided into plots of land. The owner of each plot then builds a house on his or her land. In this analogy, the house refers to a constellation's recognizable figure, while the property that surrounds the house can be thought of as each constellation's boundaries.

In 1603, only a few short years before the invention of the telescope, the astronomer Johannes Bayer created that era's most detailed atlas of the night sky, which he called *Uranometria*. He chose to identify the brightest stars in each constellation by lowercase letters from the Greek alphabet. He usually labeled a constellation's brightest star alpha, then, working his way through the traditional constellation pattern from head to toe, labeled succeeding stars beta, gamma, and so on. Once completed, he repeated the head-to-toe sequence for any fainter stars that remained, sometimes until all twenty-four letters of the Greek alphabet were used. There are many exceptions to this pattern, but it holds true for the most part. Bayer's Greek letters stuck and are still in use today. Table 1.1 lists the Greek alphabet by name and corresponding letter.

In order to extend the Greek alphabet system, the British astronomer John Flamsteed assigned numbers to all stars of about 5th magnitude and brighter in each constellation. These Flamsteed numbers begin at 1 in each constellation and increase from west to east. Fainter stars have subsequently been inventoried in other lists.

Many of the brightest stars in the night sky also have beautiful and mysterious-sounding names, such as Betelgeuse, Capella, Aldebaran, and Vega, which come to us from antiquity.

So some stars have three names. The bright star Vega is called Alpha Lyrae in Bayer's system and 3 Lyrae in Flamsteed's catalog, while nearby Albireo is also known as Beta Cygni and 6 Cygni. ("Lyrae" and "Cygni" are the genitive forms of the constellation names Lyra and Cygnus, respectively.)

Nonstellar deep-sky objects are cataloged numerically. While some of the more spectacular examples have unofficial nicknames (such as the Orion Nebula or the Andromeda Galaxy), most do not. The most famous deep-sky index of all is the Messier catalog, compiled by Charles Messier, the eighteenth-century French comet hunter, shown in Figure 1.9. Among the more celebrated members of the Messier catalog are M1, the Crab Nebula; M8, the Lagoon Nebula; M13, the Great Hercules Globular Cluster; M31, the Andromeda Galaxy; M42, the Orion Nebula; M45, the Pleiades; and M57, the Ring Nebula. The Messier catalog lists 109 of the finest nonstellar objects in the sky. Although Messier did not discover all of "his" objects (many were first spotted by his contemporary Pierre Mechain), he is credited with creating the catalog that bears his name. He numbered the objects consecutively, starting with M1, based on the order in which they were added to the catalog. (The list

*Table 1.1* **The Greek Alphabet**

| alpha | $\alpha$ | eta | $\eta$ | nu | $\nu$ | tau | $\tau$ |
|---|---|---|---|---|---|---|---|
| beta | $\beta$ | theta | $\theta$ | xi | $\xi$ | upsilon | $\upsilon$ |
| gamma | $\gamma$ | iota | $\iota$ | omicron | o | phi | $\varphi$ |
| delta | $\delta$ | kappa | $\kappa$ | pi | $\pi$ | chi | $\chi$ |
| epsilon | $\epsilon$ | lambda | $\lambda$ | rho | $\rho$ | psi | $\psi$ |
| zeta | $\zeta$ | mu | $\mu$ | sigma | $\sigma$ | omega | $\omega$ |

**Figure 1.9** *Portrait of Charles Messier (June 1730–April 1817), as painted by Desportes in March 1771. Of this portrait, Messier wrote, "This is a good likeness, except that I appear younger than I am, and I have a better expression than I have." Courtesy of Owen Gingerich.*

actually goes up to M110, but it is now generally agreed that M102 was a mistaken repeat observation of M101.)

Oddly enough, Messier assembled his listing not in an effort to record the locations of deep-sky objects, but rather to record the locations of annoying cometlike objects that hindered his comet-hunting efforts. Ironically, while all of the comets discovered by Messier have long since faded into oblivion, his catalog of 109 "nuisance" objects was what became famous, and it continues to challenge amateur astronomers to this day.

Finding all of the Messier objects is a great first observing project for new amateur astronomers. All 109 Messier objects are included in the seasonal chapters of this book. You and I are about to go Messier hunting! Now, before you say "wait a minute, I only have a small telescope," let me assure you that finding all of the Messier objects can be done with small amateur telescopes used in suburban backyards. If you live in a city, I'm afraid that you might have to travel to darker skies to see some of them, although Messier actually spotted each from a hotel rooftop in downtown Paris (but before light pollution) through telescopes no greater than 6 inches (15 cm) in aperture. Today, some amateurs complete the project with telescopes half that size.

The most comprehensive digest of deep-sky objects is the *New General Catalog of Nebulae and Clusters,* abbreviated NGC and compiled by John L. E. Dreyer in 1888. The NGC lists more than 7,800 star clusters, nebulae, and galaxies covering the entire celestial sphere. All entries are ordered by increasing right ascension. With few exceptions, all of the Messier objects are also included in the NGC. The Orion Nebula M42, for instance, is also known as NGC 1976. Dreyer subsequently assembled a pair of supplementary *Index Catalogs* (IC) that included new objects discovered after the NGC was published. Many other deep-sky inventories have been compiled since the NGC and IC listings. While the Messier, NGC, and IC listings are general compilations, most newer catalogs are segregated by object type.

### When and Where Should I Look?

The combined motions of Earth rotating on its axis and revolving about the Sun opens for us a different sky window hour by hour, as well as season by season. At the same time, the other members of the solar system are also on the move. The Moon takes about a month (actually 29.5 days) to go through its sequence of phases as it orbits Earth. Each planet orbiting the Sun can also be seen to move slowly through the sky. Mercury, the fastest, takes 88 days to orbit the Sun, and so it quickly transfers from the evening sky to the early morning sky, and back again, over a period of less than three months. At the opposite end of the scale, Pluto takes more than 248 years to revolve around the Sun. To find out which, if any, planets are visible in tonight's sky, consult Appendix B, which lists the planets' positions to the year 2015. A planetary conjunction (Figure 1.10) occurs when either two planets or a planet and the Moon appear very near each other in our sky. Conjunctions can be quite striking, whether seen through binoculars or just with the naked eye.

Beyond our solar system, each seasonal sky contains its own stars, constellations, and hidden treasures. The sky show that we enjoy on a warm summer evening, for instance, is totally different from that adorning frigid winter nights. The seasonal charts in chapters 6 through 9 will help you find out which stars will be visible in tonight's sky. Each begins with an introduction to the naked-eye sky of that season, pointing out the more prominent stars and constellations, which are also plotted on an accompanying star map. A timetable at the bottom of each map shows the hour of night depicted. For those who prefer to do their stargazing in the early morning hours instead (since that is when lower levels of light pollution often make the sky darker), each key includes times well into the early morning. A general rule to remember is that a star or constellation rises two hours earlier every month. For instance, the spring seasonal map shows the sky as it appears on June 1 at 8 P.M., as well as December 1 at 4 A.M., and several points in between. At the same time, the stars of summer are seen on May 1 at midnight.

From here, the four chapters break the seasonal skies into several detailed "sky windows." Each window is an enlarged area of the sky that plots selected deep-sky targets as well as stars to 7th magnitude. Although that's too faint to be visible to the eye alone, these stars are readily visible with binocu-

**Figure 1.10** *Mercury is but one of four planets in this photograph. On the bottom are Mercury (right) and Jupiter (left), in the middle is Mars (right) and the star Regulus (left). Finally, in between Regulus and the crescent moon is brilliant Venus. Photograph by Brian Kennedy.*

lars and finderscopes (small, low-power, spotting scopes mounted piggyback on telescopes that help the observer aim toward a target). A key map is included in each of the chapters to show where the sky windows are located with respect to each other as well as relative to the entire night sky.

### How Can I Find It?

To those new to stargazing, the naked-eye sky is tough enough to sort out. How can someone find a small, faint object buried in among all of those stars? To give you beforehand an idea of how easy or difficult an object might be to see through your telescope, I have assigned each target a "Finding Factor" to indicate just how difficult each object is to find through a small, manually operated telescope. If an object has a Finding Factor of one star, then locating it is a piece of cake. The more stars listed under the heading Finding Factor, the tougher the hunt. A target rated at a Finding Factor of five stars will probably require some fairly intense searching. Leave those until you have already bagged easier prey and developed some experience with star hopping.

"Where Am I" gives both written directions as well as matching sky windows. While some telescopes have computerized aiming systems, most readers will probably use the old tried-and-true method of star hopping. Star hopping involves going from a naked-eye jumping-off point, such as a bright star, to a faint target in a series of hops from one star or star pattern to the next. It's a great way to get to know the sky and your telescope at the same time.

Here on Earth, we are used to giving directions like left, right, up, and down, but these directions are of little use in the sky. One of the most confusing parts of viewing the sky is trying to orient star charts to agree with the stars. Which way is up? That all depends on where you are facing. Figure 1.11 shows that, depending on what direction you are facing, you will need to rotate star charts, such as those found later in this book, either clockwise, counterclockwise, or upside down. (Note that this is common to all star charts, not just those found in this book.)

To add to the confusion, most telescopes and finderscopes flip the view around, sometimes turning it upside down, other times flipping it left-to-right. Take a look at Figure 1.12 and see which view corresponds with your telescope. Remember that orientation will also depend on where you are relative to the eyepiece. If you are standing to the right side of the telescope, the view will be upside down compared to how it looks on the left side of the telescope, regardless of its design.

Once you have aligned the finder to the telescope and oriented yourself to the sky, the fun can begin. The Moon and the planets should be easy to find, if any are visible in the night sky. To help you plan your observing sessions,

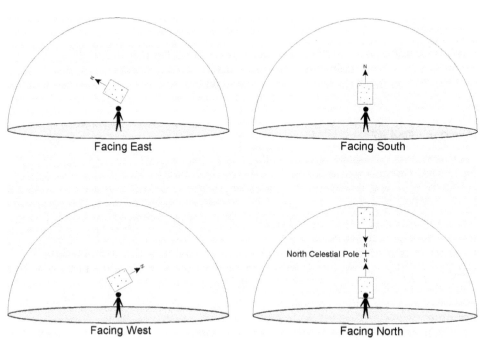

**Figure 1.11** *To use a star chart correctly, it must be turned to agree with the direction that the observer is facing.*

*The view through:*

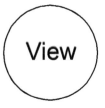  

- ● Naked eye
- ● Binoculars

- ● Telescopes without
  star diagonals
  - ★ Reflectors
  - ★ Some refractors
- ● Finderscopes

- ● Telescopes with
  star diagonals
  - ★ Refractors
  - ★ Schmidt-
    Cassegrain
    telescopes
  - ★ Maksutov-
    Cassegrain
    telescopes

**Figure 1.12** *While binoculars orient their views to agree with what our eyes actually see, all astronomical telescopes flip images around in one manner or another. Depending on the design, the image may be turned upside down or flipped left to right.*

the appendices include lists of the Moon's phases and positions of the five naked-eye planets through the year 2015.

What about locating fainter deep-sky objects? Each of the sky windows in the later chapters, as well as each object description, features a suggested star-hopping path to follow. Each star-hop casts off from a naked-eye star, constellation, or pattern, then proceeds across a portion of the sky toward the target. Along the way, star patterns such as triangles, arcs, and rectangles serve as signposts to tell you that you are on the right track. By switching back and forth between your finderscope and the sky window, you can hop from one star or star pattern to the next across the gap toward the intended target. Don't worry if you get lost along the way. Just return to the starting point and try again.

Here's an example. Let's go hunting for M1, the famous Crab Nebula in Taurus the Bull. It was Messier's first object in his list, so let's let it be ours as well. Insert your lowest-power eyepiece (remember, the longer the eyepiece's focal length, the lower its magnification), since it also has the widest field of view, which is what is needed when searching for a new target. Next, take a look at Figure 1.13. M1 is in Taurus the Bull, so the first thing to do is to find that constellation. Figure 1.13a shows M1 lies just to the west-northwest of the star Zeta Tauri, which marks one of the Bull's two horns. Step two is to aim your telescope toward that star in the sky, as shown in Figure 1.13b. Chances are your finderscope flipped the view upside down compared with what your eye alone sees. Remember to orient your finder chart to match the view.

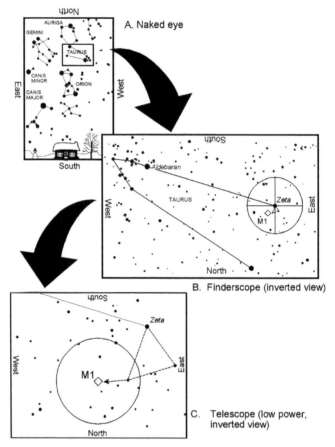

**Figure 1.13** *Star hopping is the preferred way of locating sky objects that are too faint to be seen with the eye directly. Here is the three-step approach for finding M1, the Crab Nebula in Taurus the Bull. (a) Locate the constellation or nearby naked-eye star pattern. (b) Aim your finderscope toward a known star, then star hop to fainter stars that are closer to the target. (c) Finally, look through your telescope for the target.*

Figure 1.13b is flipped around for you to illustrate the change, but you may still need to rotate it left or right to match what you are seeing exactly.

Once Zeta is centered in the finderscope, check to see that it is also in your telescope's field of view. If it isn't in there, your finderscope is probably misaligned. Double-check the finder's alignment, then try again. With Zeta in the finder, look to its north for two faint stars that join with it to form a small right triangle. See them? Good. M1 lies just to their west.

Time to switch to your telescope. Use your lowest-power eyepiece to match Figure 1.13c. Find Zeta and those two stars that form the right triangle. Slide slightly to their west. Which way is west? Look at the chart and notice the directions along its edges. Remember, the view is probably upside down, so take a moment to orient yourself. Looking through a telescope, you should see a faint smudge of light in the position marked by the diamond in the figure. That's M1. If you don't see it, move the telescope back and forth, and up and down *just a little* to see if it's nearby. If it still isn't, check the finder to make sure you have aimed it toward the correct star.

### What Does It Look Like?

I have also given each target a "WOW!" Factor to provide an indication of how spectacular an object will appear through binoculars or a telescope. Here I have chosen to rate objects not just through one, but rather through three different instruments: binoculars, small telescopes, and medium telescopes. Binoculars refer to typical 50-mm units, such as 7 × 50s and 10 × 50s. Small telescopes are defined as apertures between 3 inches (7.5 cm) and 5 inches (12.5 cm), while medium telescopes range from 6 inches (15 cm) to 8 inches (20 cm) in aperture. Telescopes larger than 8 inches (20 cm) in aperture, usually owned by intermediate and advanced amateurs, will give wonderful views of all the objects described in this book. The "WOW!" Factor uses a one-to-four-star grading system to judge an object's beauty and interest. The more stars, the more visually appealing the object. Objects noted as not resolvable are not visible through that particular instrument.

Each object is described as you would see it through binoculars and telescopes. Many new stargazers are disheartened to learn that their telescopes will not show sky objects as they are depicted in the brilliant color photographs that often adorn the outsides of their telescope boxes. Sorry, that is the sad reality. But just because the view doesn't look like the backdrop of a space science-fiction movie doesn't mean that astronomy is not for you. Quite the contrary. You will find, as I did over the years, that what appears to be a vague, formless mass at first will often transform into a striking sight after concentrated study. So, take heart. If an object appears uninteresting at first, give it some time. If you study each object carefully, using some of the tricks and techniques described in this book, then you will be amazed at what you can see. Best of all, you are seeing it live through your own telescope or binoculars, not in some sanitized photograph. That thrill and sense of accomplishment can never be captured in a photograph.

All observers have had those moments, however, when, even after great care has been taken to aim toward an intended target, it simply isn't there. No matter how often you go back and forth to one of the charts, the object just isn't there. It can drive you nuts! But just because something isn't there at first pass doesn't mean that it is time to move on to another object. Here are a few secrets that may help render an invisible object visible.

Many observers overlook the simple need to let their eyes become accustomed to darkness before searching for faint objects. While most people's eyes partially adapt in about 20 to 30 minutes, the entire process can take an hour or more to complete. Plan your night's observing program so that brighter objects are viewed first. Wait at least an hour before you begin to search out those faint fuzzies that are just on the brink of visibility.

Next comes patience. Take a deep breath and slow down! Just because an object doesn't reveal itself immediately doesn't mean that it won't after a few minutes of concentrated searching. If eye fatigue sets in, move away from the eyepiece and take a short break.

If a faint object refuses to be seen when stared at directly, try looking at it with your peripheral vision. This technique, called *averted vision,* can be very successful for spotting deep-sky objects. With peripheral vision (that is, looking to one side or the other of the target being observed), light falls on a more sensitive part of the eye's retina. Frequently, objects that are invisible when viewed directly will be seen with averted vision.

Of course, even averted vision won't make much difference if the sky isn't clear and dark. Nothing short of clouds can ruin the sky more than *light pollution,* the curse of the modern astronomer!

There are two kinds of light pollution: sky glow and local light pollution. Sky glow is general light pollution that comes from buildings, streetlights, roadside billboards, floodlights, and civilization in general. It can turn a beautifully clear day into a soupy, grayish night. Can anything be done to combat sky glow? The good news is yes, but progress is slow. In an attempt to recapture one of our greatest natural wonders, the night sky, some municipalities have enacted legislation against overlighting. Connecticut became the first state to pass statewide parameters restricting nighttime lighting. Similar legislation has been advanced in other states as well. Will these light-pollution laws make a difference in our view of the night sky? Only time will tell, but it is a step in the right direction.

Localized light pollution, such as from a streetlight or porch light, is a little easier to deal with. To help combat this distraction, use a personal cloaking device, which is nothing more than a piece of dark cloth placed over your head and eyepiece, like an old-time photographer's shroud. The only drawback to the cloaking device is that in damp weather it tends to accelerate eyepiece fogging.

### What Am I Looking At?

Astronomers have been asking this all-important question since the dawn of time. What exactly are we looking at when we view these objects? Each of the chapters to come includes a discussion of the latest findings about each target. But it seems at times that the more we learn about these celestial objects, the more questions we need to ask. That is perhaps the most compelling part of astronomy.

The night sky holds a lifetime's worth of fascination for stargazers. Whether you own binoculars or a telescope, or if you live in the city, the suburbs, or the country, there is always something new to look at. Before dashing off to your telescope, however, review the chapters to come, especially the one that covers the night sky as it will appear tonight. Take a look at the sky objects on tonight's menu and start with the brighter offerings. If the Moon or a planet is visible, that's an excellent beginning. Then, work your way through the deep-sky objects. Take your time and enjoy the view.

# 2

# The Moon

Measuring 2,173 miles (3,476 km) in diameter and located an average of 238,800 miles (384,500 km) from Earth, the Moon is our closest neighbor in space. Its harsh beauty through binoculars or a telescope is an incredible sight. Dark maria, towering mountains, and all those craters appear so close, you almost feel you can reach out and touch them. No other celestial object shows as much exquisite detail as the Moon, and so it deserves a special place in any book that discusses the night sky.

Humans had long dreamed of traveling to the Moon. What seemed like an impossible goal was reached in July 1969, when *Apollo 11* astronauts Neil Armstrong and Edwin Aldrin touched down on a flat plain called Mare Tranquilitatis (the Sea of Tranquility). Between then and December 1972, a dozen astronauts visited the Moon, photographed its "magnificent desolation," as Aldrin put it, left behind measuring equipment, such as seismometers, and returned with more than 800 pounds of lunar samples. From these, as well as from the more recent robotic Clementine and Lunar Prospector missions, we now have a pretty good picture of what makes our nearest neighbor in space tick.

Most of the samples returned by the Apollo missions show a surface that is largely made up of silicates, like Earth's crust and mantle. The dark areas, called *maria* (pronounced MAR-ay-uh; singular: *mare*, MAR-ay) or seas, were once believed to be bodies of water, but they are actually composed of basalts, dark material that is similar to the lava that erupts from many earthly volcanoes. The mountainous lunar highlands that rise above and surround the maria are made primarily of anorthosite, a comparatively light-colored, low-density rock that solidified as the Moon cooled after its formation some 4.1 to 4.5 billion years ago.

From this, we can finally answer with some authority the question "where did the Moon come from?" It is now generally believed that in the early history of the solar system, Earth was a victim of a giant impact by another, rogue object, which struck Earth at a glancing angle. In the process, the shattered rubble from Earth, comprised largely of material from our planet's crust, was ejected into space. Eventually, that material coalesced to form the Moon.

## Craters

Immediately after its formation, the Moon was both very active volcanically and constantly being bombarded by remnant debris from the formation of the solar system that still floated among the planets. An initial period of heavy cratering took place some 4 billion years ago, forming several huge impact basins. Though the heavy bombardment eventually ceased, the lunar vulcanism flourished, eventually flooding many of the craters and deep basins with lava. Between 3.3 and 3.8 billion years ago, as the inner solar system cooled from the forces that created it, lunar vulcanism slowed, then ceased. The Moon's core cooled and solidified. The flooded impact basins, which had been filled to their brims with lava, also solidified to form the maria. A period of less intense cratering continued after the Moon cooled. As leftover debris from the solar system's formation eventually dispersed—by either impacting the planets or being sucked into the Sun—the era of lunar cratering ended, leaving us the scarred surface that we marvel at today.

### Moon Phases

Probably one of the first things you learned about the Moon in early childhood was that it changes shape. These *phases* are caused by sunlight sweeping across the lunar surface as the Moon orbits Earth. Figure 2.1 shows the progression of the phases through one complete cycle. The four major phases—New Moon, First Quarter, Full Moon, and Last (or Third) Quarter—are separated by about a week and often marked on calendars, while the intermediate phases—crescent and gibbous—bridge the gaps. The words *waxing* and *waning* are used to describe the crescent and gibbous phases further. *Waxing* means that the Moon appears to be growing in size night after night, which it does as it goes from New Moon to Full, while *waning* means that it appears to be shrinking as time goes on, as it does returning from Full back to New. The waxing phases are those seen in the early evening, while waning phases are usually seen late at night and in the early morning.

The Moon takes 27 days 7 hours 30 minutes to orbit Earth, but you may notice that there are more than 29 days (29 days 12 hours 44 minutes, to be exact) between like phases (e.g., from one New Moon to the next). This 2-day difference is caused by the fact that as the Moon orbits Earth, Earth continues in its orbit of the Sun. As a result, the Moon needs to travel an extra 2 days to catch up with the Sun, as seen from our perspective.

Each orbit is referred to as a *lunation* and is measured in terms of days. Day 0 is marked by the New phase, which is invisible, since it is so closely aligned with the Sun in our sky (if they are aligned exactly, a solar eclipse occurs; see chapter 4). Keeping with this convention, a 4-day-old Moon signifies the Moon's "age" of 4 days after New Moon, and so on. Using this approach, we typically refer to the First Quarter Moon as being 7 days old. The Full Moon is also thought of as the 14-day-old Moon, while the Last Quarter is 21 days old. The dates of the four major lunar phases from 2003 to 2015 are available on my Web site, www.philharrington.net.

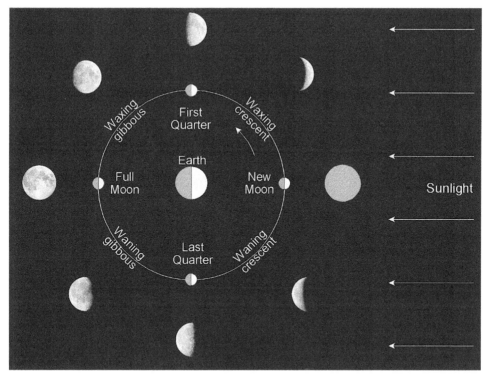

**Figure 2.1** *Phases of the Moon during a complete orbit of Earth, from a perspective in space. Although half of the Moon is always lit by the Sun, only a portion of the sunlit face is usually visible from Earth. The images around the outside edge of the diagram show what an observer would see from Earth.*

## Moon Gazing

FINDING FACTOR:    *

"WOW!" FACTOR:    Binoculars: ****        Small telescopes (3″ to 5″): ****
Medium telescopes (6″ to 8″): ****

### What You'll See

**With the Naked Eye.** While the bulk of this chapter concentrates on specific lunar landmarks visible through telescopes and binoculars, we begin our visit with just our eyes. Let's start with the crescent phases, which are striking to the naked eye, hanging in the darkening sky right after sunset or in the east immediately before dawn. Can you see the "young Moon in the old Moon's arms?" That's a phrase sometimes used to refer to *earthshine*. Earthshine (sunlight reflecting off Earth) causes the dark portion of the Moon to appear faintly illuminated. When we see the crescent Moon in our sky, an observer on the Moon would see a gibbous Earth. Earth would appear so bright and large in the lunar sky that it would bathe the lunar landscape in reflected sunlight. A similar effect is seen here on Earth at and around the Full Moon phase. Light from the Moon floods the night, making it possible to see things that would otherwise be invisible in the darkness. (That effect is called, appropriately, *moonshine*, and was the light used by bootleggers to run stills and make illegal liquor during the days of Prohibition. Hence, the term moonshine for liquor.) While the Full Moon is a beautiful sight in the sky, its intense light all but obliterates nearly every other sky object. As a result, the nights around Full Moon are least favorable for stargazing and, in fact, moongazing, as well.

Because the Moon isn't perfectly round but is rather somewhat pear shaped, it is gravitationally locked in its orbit of Earth, with one side constantly facing our planet and the other side facing away. As a result, when we look at the Moon from Earth, we are only seeing half the story. The far side of the Moon remains perpetually turned away, revealing itself only to spacecraft in lunar orbit.

Actually, that is not completely true. It turns out that the Moon appears to wobble a bit as a result of variations in the Moon's orbit motion as well as the inclination of the Moon's equator with respect to its orbital plane. This lets us see about 59 percent of the Moon's disk from Earth. This effect is called *lunar libration*.

**Through Binoculars and Telescopes.** Just as we plan our vacations here on Earth based on the season and sights to see, so too should we plan our lunar excursions based on the Moon's phase. Some phases are better for sightseeing than others. Our tour begins with the waxing phases immediately after New Moon, continues through Full Moon, and back through the waning phases.

Let's first settle the matter of lunar directions. If we were standing on the Moon reading this book, we would reckon the Sun rising in the east, moving very slowly across the sky (recall how a lunar day takes more than 29 Earth days to complete), and setting in the west. East is east and west is west, which seems to make perfect sense. Prior to the early 1960s, however, most books used a geocentric (Earth-based) frame of reference. In this old system, the western edge, or *limb*, of the Moon was closest to *our* western horizon. Likewise, the eastern limb was nearest our eastern horizon. North and south remain unchanged as well.

But with the advent of the space program, a new, more logical system emerged. When we now refer to one crater being to the east of another, we are talking about *lunar east*, not terrestrial east. Likewise, west refers to *lunar west*, which is the edge toward our eastern horizon. Observing the Moon when it is on the meridian from the Northern Hemisphere here on Earth, the eastern edge of the Moon lies to the right, the western edge to the left. Sorry if this is a bit confusing, but I think you will see, with time, that it does make sense.

Of course, most telescopes flip all this around anyway, either upside down (north-south) or mirror imaged (east-west), or possibly both!

### Waxing Crescent: Days 1 through 6

Trying to spot the Moon one night after its New phase is difficult at best. By the time another 24 hours have passed, a thin sliver of moonlight can be spotted hanging low in the western sky as twilight replaces day. While an aesthetically beautiful sight, little or no detail can be spotted for these first couple of nights after the New phase. By Day 3 and Day 4 (Figure 2.2), however, the Sun has risen high enough above the lunar horizon to reveal the first surface features.

The first lunar mare to come into view after New Moon is **Mare Crisium,** the Sea of Crises. This large oval plain measures 270 miles by 350 miles (435 km by 560 km), with the long dimension running east to west. Through a telescope, however, Mare Crisium appears to be wider north-south than east-west. This is an illusion caused by foreshortening resulting from the Moon's curvature.

**Mare Fecunditatis,** the Sea of Fertility, can be seen to the south of Mare Crisium. Only the twin craters **Messier** and **Messier A** (referred to in some older books as Pickering) blemish this mare's smooth, dark surface. These two craters are fascinating to watch night after night, as the Sun rises higher in their sky. Take an especially careful look at Messier A. Can you see what looks like a double comet's tail extending to the crater's east? This tail is actually a pair of lunar rays, formed from material that was ejected when the meteoroid that caused the crater impacted, apparently at a narrow angle from the west. Watch as the rays continue to grow in intensity as the Moon phases from crescent through gibbous to Full. Also track the lighting changes to the crater **Langrenus,** which lies on the southeastern shore of Mare Fecunditatis. As the Sun rises higher in the Moon's sky, 82-mile-wide (132-km-wide) Langrenus seems to almost catch fire as its floor is transformed from a dull gray to a brilliant whitish glow.

Several notable craters are visible to the north of Mare Crisium during the waxing-crescent phases. Almost touching the sea's "coastline," **Cleomedes** measures 78 miles (126 km) across, with sheer walls that descend nearly 3 miles (5 km). Switching to a higher magnification, try to spot a thin, sinuous channel, or *rille,* that runs for about 20 miles (32 km) along the crater's floor. The crater **Burckhardt** is just to the north of Cleomedes, while sharp-walled **Geminus** is farther north still. Finally, to the west of Mare Crisium is the tiny crater **Proclus.** Though it may not look like much now, come back in a few nights and watch as its striking ray pattern catches light from the rising Sun.

**Atlas** and **Hercules** make up a formidable pair of craters to the northwest of Mare Crisium. Atlas features a system of rilles along its 10,000-foot-deep (3,050-meter) floor, while Hercules has sharper walls as well as a distinctive, dark floor.

Finally, the **Rheita Valley** first sees sunrise around Day 4 of every lunation. You'll find more information about it and the nearby crater Rheita in the next section.

### First Quarter: Days 5 through 8

The later waxing-crescent through First Quarter (Day 7) and early waxing-gibbous phases (Figure 2.3) display a tremendous variety of lunar terrain. Dominating the equatorial zone are the vast expanses of four lunar seas. To their north are many scattered large craters, while to their south lies the coarse beauty of the Moon's south polar region, where there are so many craters that it is often difficult to distinguish one from another.

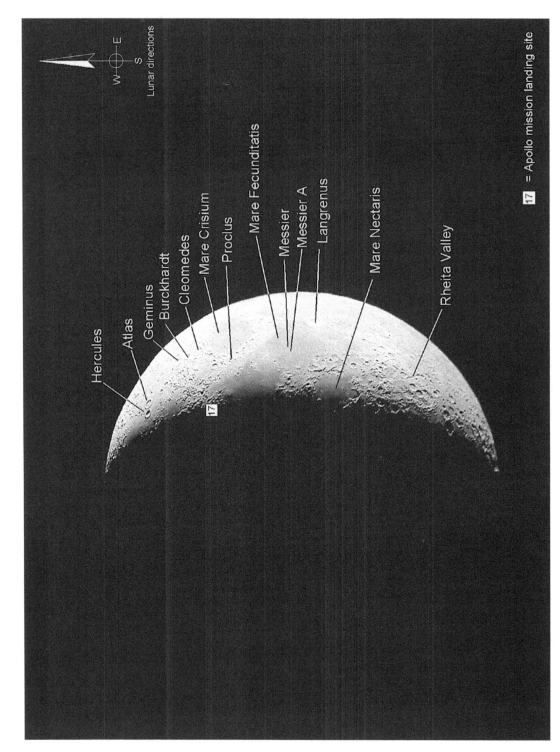

**Figure 2.2** *Four-day-old Moon. Photograph taken by the author through an 8-inch f/7 reflector and a 26-mm eyepiece.*

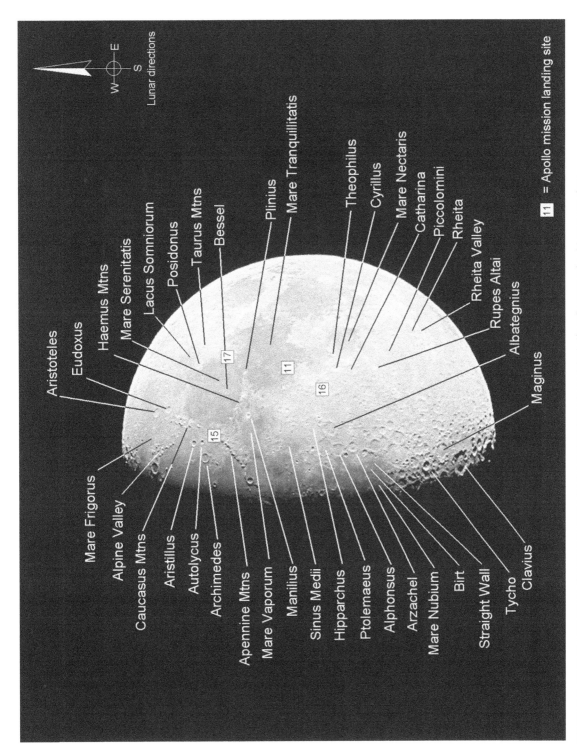

**Figure 2.3** *Eight-day-old Moon. Photograph taken by the author through an 8-inch f/7 reflector and a 26-mm eyepiece.*

Let's begin with the maria. In addition to Mare Crisium and Mare Fecunditatis, which first came into view a few nights after New Moon, **Mare Nectaris** (the Sea of Nectar), **Mare Serenitatis** (the Sea of Serenity), **Mare Frigoris** (the Sea of Cold), and the most famous of all, **Mare Tranquilitatis** (the Sea of Tranquility) are now visible. Can you see any "waves" in the Sea of Tranquility? Actually, those aren't waves but wrinkles that resulted when rapid cooling of the Moon caused the flowing lava to stop dead in its tracks.

A trio of easily seen craters can be found just "inland" from the western edge of Mare Nectaris. **Theophilus, Cyrillus,** and **Catharina** are all large enough to be visible in steadily held binoculars, and telescopes reveal them easily. Though similar in diameter, all three look quite different. Theophilus is the youngest, as we can see from its sharp edges and the fact that it partially overlaps neighboring Cyrillus. It also shows a more prominent central peak than either of the others, another indicator of relative youth.

From Catharina, follow the curve of **Rupes Altai,** a remnant wall formed from the impact that created Mare Nectaris, toward the southeast. Notice how Rupes Altai runs parallel to the edge of the mare. The southernmost tip of Rupes Altai lies near another prominent crater, **Piccolomini.** Measuring 55 miles (89 km) in diameter, this sharp-walled crater has a strong central peak that stands out especially well just after sunrise some five days after New Moon.

Continue past Piccolomini, moving toward the limb, to find the crater **Rheita** as well as the unusual **Rheita Valley.** At about 300 miles (500 km) long, the Rheita Valley is the longest formation of its kind on Earth-facing side of the Moon. It seems to radiate from the center of Mare Nectaris and likely shares a common origin. Again, the best time to view Rheita and the Rheita Valley is about two or three nights before First Quarter.

The Sea of Tranquility was the site where Neil Armstrong became the first human in history to set foot upon another world, speaking those immortal words on July 20, 1969, "That's one small step for [a] man, one giant leap for mankind." Today, more than thirty years after that historic event, we wonder when we will return to our neighbor in space. Armstrong and lunar module pilot Edwin "Buzz" Aldrin set their craft down southwest of the sea's center, marked by the boxed "11" in Figure 2.3. The best time to zero in on the landing site is around Days 6 and 7 (First Quarter), when the curtain of daylight first opens on the area.

Some authorities believe that the Sea of Serenity is the oldest lunar mare of all. Covering an area 360 miles by 420 miles (580 km by 675 km), it is bound by the **Haemus Mountains** to the south, the **Apennines** to the west, the **Caucasus Mountains** to the north, and the **Taurus Mountains** to the east. No mission from Earth has ever touched its nearly crater-free surface. Look carefully and you will find a few impact craters within this otherwise serene sea. Its most conspicuous crater is **Bessel,** found south of center. Bessel is surrounded by a small, but pronounced ejecta blanket. Another obvious crater, **Plinius,** stands as an island where the two maria join. **Posidonus,** a broad crater (technically referred to as a walled plain), lies along the northeastern shore of Mare Serenitatis, near **Lacus Somniorum,** the Lake of Dreams.

Another prominent pair of craters lies to the north of Mare Serenitatis: **Aristoteles** and **Eudoxus.** Aristoteles is a conspicuous 54-mile-diameter (87-km) crater found on the southern edge of Mare Frigoris. The crater's steeply banked, terraced walls plummet 12,000 feet (3,700 meters). Eudoxus measures 42 miles (67 km) across with terraced walls that plunge 14,500 feet (4,400 meters). The latter is located just east of the Caucacus Mountains, which partially bridge the strait between Mare Serenitatis to the east and Mare Imbrium to the west (still in shadow). The highest peaks in the Caucacuses rise to 20,000 feet (6,000 meters) above sea level and are an especially impressive sight when sunlight first strikes their summits.

To their west, directly on the terminator (the sunrise line) at First Quarter, a deep wound in the lunar surface is seeing sunshine for the first time this lunation. This is the **Alpine Valley,** which stretches for about 95 miles (153 km), oriented more or less southwest-northeast.

The most striking mountain range on the Moon, the Apennines, reaches up to meet the Caucacus Mountains from the south. Try to catch them just as the Sun rises there in the morning or as it sets in their evening. The low-angle lighting really brings out the three-dimensional relief effect. It's especially fun to watch as sunlight first strikes the summit of an individual peak, creating a floating speck of light. Then, as the Sun rises higher in the sky, the light slowly slides down the mountainside until it finally floods the adjacent valleys. One of the Apennines's most prominent peaks, Mount Hadley, rises 14,800 feet (4,500 meters). Look for it in the northwestern portion of the range. After the Sun has risen fully above it, take a look for a small, sinuous channel, or rille, to its west called Hadley Rille. It was here that the *Apollo 15* astronauts spent nearly three days exploring the Moon, using the first lunar rover.

Several prominent craters are revealed each night around the First Quarter phase. Moving from north to south, our first stop is the triangle formed by **Aristillus, Autolycus,** and **Archimedes.** Aristillus is a rayed crater that spans 34 miles (55 km) and contains a prominent three-summit central peak, while smaller Autolycus measures 24 miles (39 km) across. Of the three, dawn comes last to Archimedes, which stretches for 51 miles (82 km) across. Long after it was formed, Archimedes was flooded by hot lava that broke over the crater walls and flooded its floor, entombing forever whatever lay below.

Go out on the night of First Quarter and look at the center of the terminator. There, you will see a grayish, circular formation known as **Sinus Medii** (Central Bay), a small mare that is exactly in the middle of the side of the Moon facing Earth. Just to its north, you'll see sunrise over **Mare Vaporum,** the Sea of Vapors. The most prominent crater in the vicinity is **Manilius,** on the sea's western edge. As the Sun rises higher in the sky, watch the crater's ejecta field brighten. Sharp eyes might also spot the Hyginus Rille crossing the southern portion of the sea. It appears as little more than a thin pencil line but in reality measures 140 miles (225 km) long by 2 miles (3 km) wide.

Slowly travel south of Sinus Medii, where a number of large craters greet you along the way to the south polar region. The large crater nearest to Sinus Medii is **Hipparchus,** an old feature whose walls have been battered by more recent impacts, such as the one that created **Albategnius,** which lies just to the south of Hipparchus. The craters **Ptolemaeus, Alphonsus,** and **Arzachel** come into light on Day 8, forming a prominent north-south line of craters, almost touching one another's borders to the east and south of Albategnius.

The 8-day-old Moon is, by far, my favorite phase of the entire lunar cycle, for it marks the appearance of three impressive craters. Leader of this pack is mighty **Clavius,** second largest crater on the side of the Moon facing Earth. Under the right conditions, sunlight can catch just the top of its huge, 132-mile-by-152-mile (213-km-by-245-km) rim, causing a bright ring to protrude into the cold, dark lunar night. On the floor of Clavius are several smaller impact craters, with **Porter** on its northern rim the most obvious. Just north of Clavius is the crater **Tycho.** Its sharp walls and conspicuous central peak really set it apart from the neighboring features. While impressive even in this early phase, Tycho really asserts itself as one of the Moon's most spectacular craters during the gibbous and Full phases. Once Clavius and Tycho are identified, try your luck with **Maginus,** to the northeast of Clavius.

Slide back northward along the terminator on Day 8, toward **Mare Nubium,** the Sea of Clouds, which is just coming into sunlight. There you will also see a small crater named **Birt** and, a bit closer to the shoreline, a black line that looks almost too straight to be natural. This line, which looks almost like an incision in the Moon, is called the **Straight Wall,** a sheer cliff that seems to rise out of nowhere. The walls of this 75-mile-long (120-km) cliff suddenly climb at a 40° angle to a height of 1,200 feet (366 meters). On the Moon's eighth day, you are not looking at the wall itself but rather at its darkened face which has yet to be lit by sunlight. The cliff's shadow is still being cast onto the lower "sea floor." Come back in two weeks, on the night after Last Quarter, and you will see the Sun's light fully illuminate the then-shadowless cliff, which appears gleaming white.

### Waxing Gibbous: Days 9 through 13

With the Moon growing larger and brighter, the challenge facing observers during the gibbous phases (Figure 2.4) is balancing image intensity with viewing comfort. I strongly recommend that you purchase a neutral-density Moon filter that screws into your eyepiece. Available from many different companies, these filters dim the Moon's brightness to a comfortable level, making it easier on the eyes to see fine details without being dazzled.

As the terminator continues to push across the lunar west it slowly uncovers many fascinating features. None is more prominent than the crater **Copernicus.** Measuring 60 miles (100 km) in diameter, Copernicus is easily found at the point where **Mare Imbrium** (the Sea of Rains) meets **Oceanus Procellarum** (the Ocean of Storms). When the Moon is 9 days old, watch as sunlight first bathes the crater's sharply defined walls with light, catching the strong central mountain peak before sliding down to the crater floor. Also take a look around the outer edge of Copernicus, where the rugged, textured surface tells of the powerful impact that formed the crater billions of years ago.

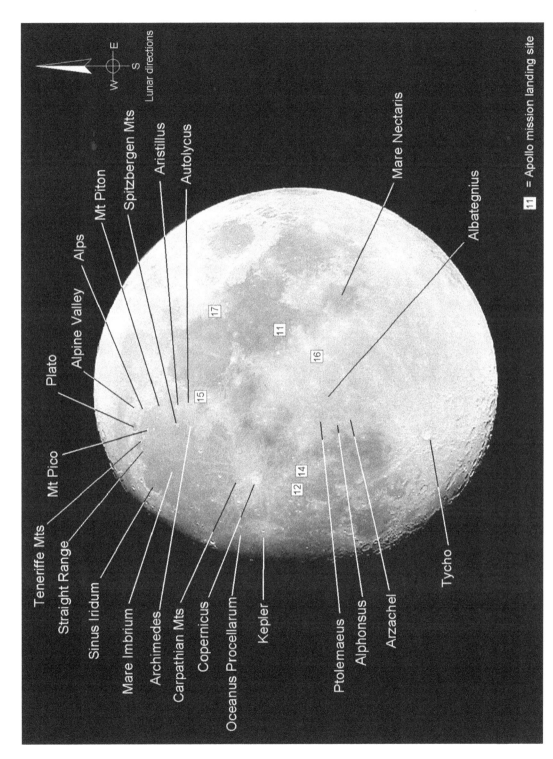

**Figure 2.4** *Eleven-day-old Moon. Photograph taken by the author through an 8-inch f/7 reflector and a 26-mm eyepiece.*

Mark your calendar to come back in a few nights, closer to Full Moon, when the brilliant ray system of Copernicus explodes into view against the darker background of the maria. Its starburst pattern is unmistakable through the most modest binoculars and can even be seen with the eye alone.

Just north of Copernicus, the **Carpathian Mountains** rise as high as 6,600 feet (2,000 meters) above lunar sea level. These mark a portion of the edge of Mare Imbrium, dividing it from Oceanus Procellarum, the largest of the lunar plains. Dominating the waxing gibbous phases, the Ocean of Storms covers more than one million square miles and was also the site of the *Apollo 12* lunar landing mission in December 1969.

Two nights after Copernicus comes into view, a smaller crater to its east sees dawn. This is **Kepler.** Although only 20 miles (32 km) across, Kepler is one of the most prominent craters on the Moon once the Sun rises high in the Moon's sky and lights the crater's bright ray pattern. Because they are completely isolated from the bright highlands regions, the rays from both Kepler and Copernicus really stand out as we approach Full Moon.

Set sail across Mare Imbrium for its northern shore, where you can find safe harbor in **Sinus Iridum,** the Bay of Rainbows. Sinus Iridum is actually a crater, or more correctly, what's left of a crater. Sometime after the impact that created the crater, the crater's northern wall was breached by a flood of hot lava welling up from the nearby mare. The southern rim of the crater remains "buried at sea" under solidified lava.

From Sinus Iridum, follow the thin strip of rugged lunar terrain that separates Mare Imbrium from Mare Frigoris (the Sea of Cold) to the northeast. There, you will find the large, dark-floored crater **Plato.** Its floor, flooded with mare basalts, is the darkest of any crater found on the Moon's Earth-facing side and appears even darker than neighboring Mare Imbrium and Mare Frigoris. This flooding must have occurred relatively late in the era of lunar vulcanism, since the crater floor is spoiled by only a few small craterlets. In fact, trying to see all five of the tiny craters within Plato is a fun test of both an observer's eyesight as well as a telescope's mettle.

The western edge of Plato marks the eastern boundary of the lunar **Alps.** Like their earthly counterparts, the Alps are an impressive mountain range, stretching for 350 miles (560 km) from Mare Frigoris in the north along the perimeter of Mare Imbrium. The highest peak in the lunar Alps has an altitude of 14,000 feet (4,300 meters). When viewed during the waning gibbous phases, the striking crater Plato will help point your way to these slopes. Be sure to take a peak at the **Alpine Valley,** a gap that slices straight through the mountains, acting as a natural "canal" connecting the two waterless lunar seas.

Several mountain peaks rise from Mare Imbrium, just south of Plato. These include the **Teneriffe Mountains,** the **Straight Range,** and **Mount Pico.** Another, **Mount Piton,** lies due south of the Alpine Valley. Although the surrounding valley was flooded by molten rock during the Moon's volcanic period more than three billion years ago, these peaks were tall enough to remain above the rising lava level. All serve as testimony that the Moon was very different before the maria were formed.

Mare Imbrium is home to three prominent craters that see sunrise during the waxing gibbous phases. Largest of the three is **Archimedes.** Like Plato,

## Apollo Missions

Between July 1969 and December 1972, six teams of United States astronauts ventured across the gap between Earth and the Moon to land and walk on our only natural satellite. During each of those missions, they collected soil and rock samples, set up instruments for measuring moonquakes and determining the exact distance to the Moon, and planted the U.S. flag.

Here is a capsule summary of each of those missions, along with tips for spotting each of the landing sites. No, you can't see the flags that were left behind, but you can see the rugged terrain that each mission faced when trying to set down their lunar module.

*Apollo 11*
Dates: July 16–24, 1969
Landing site: Sea of Tranquility
Best phase to view landing site: Waxing crescent
Astronauts: Neil Armstrong (commander), Edwin Aldrin (lunar module pilot), Michael Collins (command module pilot)
Accomplishments: First two astronauts to walk on the Moon; planted U.S. flag; conducted limited scientific experiments; left plaque that stated, "We came in peace for all mankind"; returned with lunar soil samples.

*Apollo 12*
Dates: November 14–24, 1969
Landing site: Ocean of Storms
Best phase to view landing site: Waxing gibbous
Astronauts: Charles Conrad (commander), Alan Bean (lunar module pilot), Richard Gordon (command module pilot)
Accomplishments: Landed within walking distance of *Surveyor III*, an unmanned lunar probe that landed in April 1967; returned with parts of the *Surveyor* spacecraft to study effects of the Moon's harsh environment on metals.

*Apollo 14*
Dates: January 31–February 9, 1971
Landing site: Fra Mauro crater
Best phase to view landing site: Waxing gibbous

Astronauts: Alan Shepard (commander), Edgar Mitchell (lunar module pilot), Stuart Roosa (command module pilot)
Accomplishments: Returned with lunar soil samples; Shepard hit golf balls with an improvised golf club.

*Apollo 15*
Dates: July 26–August 7, 1971
Landing site: Hadley Rille, near the Apennine Mountains
Best phase to view landing site: First Quarter
Astronauts: David Scott (commander), James Irwin (lunar module pilot), Alfred Worden (command module pilot)
Accomplishments: First mission to bring along the lunar rover "moon buggy," enabling astronauts to travel farther from landing site; found 4.5-billion-year-old "Genesis rock" that confirmed scientists' estimate of the age of the solar system.

*Apollo 16*
Dates: April 16–27, 1972
Landing site: Crater Descartes
Best phase to view landing site: Waxing crescent
Astronauts: John Young (commander), Charles Duke (lunar module pilot), T. Kenneth Mattingly (command module pilot)
Accomplishments: Using lunar rover, gathered lunar soil samples that proved the Moon was once volcanically active.

*Apollo 17*
Dates: December 7–19, 1972
Landing site: Adjacent to the craters Taurus and Littrow
Best phase to view landing site: Waxing crescent
Astronauts: Eugene Cernan (commander), Harrison Schmidt (lunar module pilot), Ronald Evans (command module pilot)
Accomplishments: Using lunar rover, explored northeastern quadrant of the Moon; Schmidt became the first scientist to visit the Moon; left plaque on the Moon that read, "Here man completed his first explorations of the Moon. May the spirit of peace in which we came be reflected in the lives of all mankind."

Archimedes was flooded by lava during the Moon's volcanic era. **Aristillus,** to the northwest of Archimedes, is the freshest crater of the three, with a prominent central peak visible. As the Sun rises higher in the lunar sky, a system of bright rays appears to spray from the crater into the surrounding plains, making Aristillus easy to find. Just north of Archimedes and east of Aristillus lie the **Spitzbergen Mountains,** another isolated cluster of peaks rising from the smooth floor of Mare Imbrium.

Moving northward, cross Mare Cognitum ("the sea that has become known") and continue toward Mare Nubium, site of the Straight Wall, detailed previously but still visible. The craters Hipparchus, Albategnius, Ptolemaeus, Alphonsus, and Arzachel are also still visible to the west of Mare Nubium.

### Full Moon: Day 14

While many people are under the impression that the Moon is full for several nights, in reality it attains that phase at a single, precise moment in each lunation. Full Moon (Figure 2.5) is the point in the Moon's orbit when it appears directly opposite the Sun in our sky. In more precise, celestial-coordinate terms, Full Moon is that instant when the Moon is separated from the Sun by exactly 12 hours 0 minutes in right ascension. Of course, if the Moon is *precisely* opposite the Sun in our sky, then we will witness a lunar eclipse. Those magnificent events will be spoken about later.

Many are also under the false impression that the Full Moon is the best phase for lunar viewing. This would certainly seem to make sense at first, since the entire Earth-facing side is illuminated, but in truth, the Moon appears overly bright and very stark when full. The shadows along the terminator, which give such wonderful three-dimensional relief throughout the phases leading to Full, are gone. Instead, sunlight shining directly on the lunar surface makes everything look flat.

While shadowing is absent, the direct lighting accentuates the many bright features scattered across the lunar surface, such as crater rays and ejecta fields. Certainly, the single most impressive ray system emanates from the crater **Tycho,** which we met at sunrise on Day 8. Back then, Tycho was overshadowed by its neighbor Clavius, but now it has really caught the Sun's light to create a magnificent starburst pattern near the Moon's south pole that is even visible to the naked eye. The 52-mile-diameter (84-km) crater itself gleams a radiant white, surrounded by a slightly darker area about 90 miles (145 km) across. But what's most impressive is its huge system of rays spraying radially outward for more than 900 miles (1,450 km). The rays extend far into Mare Nubium, Mare Nectaris, and even across Mare Serenitatis, as well as over to the far side beyond the Moon's south pole. Studies reveal that Tycho is likely the youngest of the Moon's prominent craters, having been created about 100 million years ago.

**Copernicus** also puts on a bright show during the Moon's fullest phases. The crater itself appears a brilliant white, while a slightly dimmer ejecta field

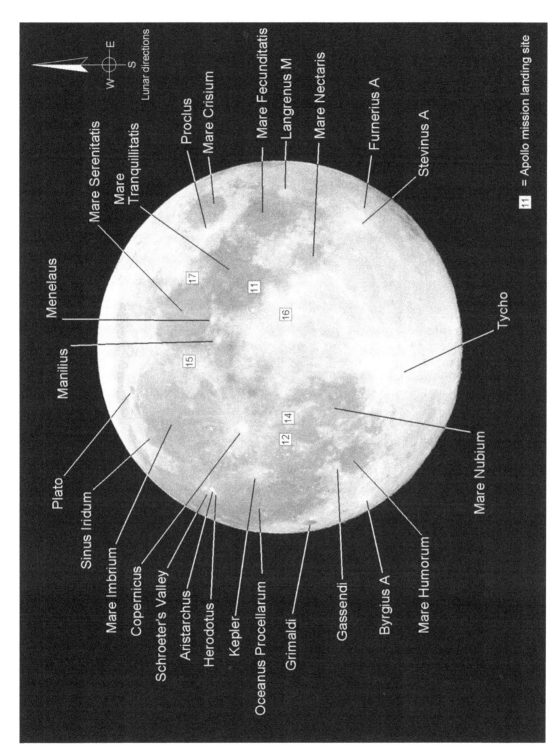

Plato  Menelaus
Sinus Iridum  Manilius  Mare Serenitatis
Mare Imbrium  Mare Tranquillitatis
Copernicus  Proclus
Schroeter's Valley  Mare Crisium
Aristarchus  Mare Fecunditatis
Herodotus  Langrenus M
Kepler  Mare Nectaris
Oceanus Procellarum
Grimaldi  Furnerius A
Gassendi  Stevinus A
Byrgius A
Mare Humorum  Tycho
Mare Nubium

Lunar directions

11 = Apollo mission landing site

**Figure 2.5** *Fourteen-day-old Moon. Photograph taken by the author through an 8-inch f/7 reflector and a 26-mm eyepiece.*

scatters rays into Oceanus Procellarum and Mare Imbrium. To its west, **Kepler** looks like a miniature Copernicus, with its ray system scattering into the Ocean of Storms as well. Finally, the ray system from the crater **Aristarchus** looks like a lighthouse adrift in the middle of the same ocean, to the northwest of Kepler and Copernicus. In fact, Aristarchus is so bright that William Herschel, arguably the greatest observational astronomer of all time, thought at first that it was an erupting volcano! A less well-defined crater, **Herodotus,** lies just to the west.

Once Aristarchus is in view, switch to a high-power eyepiece (say, around 200×) and look for a thin, twisting line winding its way just to the northwest. This is called Schroeter's Valley, although technically it is not a valley but rather a rille, cut into the Moon's surface by hot lava billions of years ago. Schroeter's Valley is the largest feature of its type on the Moon, about 100 miles (160 km) in length.

Just about directly on the western edge, or limb, of the Full Moon is **Grimaldi,** a flooded impact basin that looks just like a miniature Mare Crisium. Formed from a giant impact, then subsequently flooded with hot lava, Grimaldi shows a perfectly smooth surface through amateur telescopes. Its elliptical appearance is caused by the Moon's curvature. Grimaldi is in fact nearly circular at 140 miles by 145 miles (225 km by 233 km) across.

Moving southward along the lunar limb from Grimaldi, you'll come to a bright ejecta field surrounding a small impact crater. That crater, which went completely unnoticed when it was on the terminator, is **Byrgius A.** The "A" designation indicates that this crater overlaps the wall of a larger crater for which it is named (in this case, Byrgius). But that larger crater is completely overwhelmed by the bright rays emanating from the smaller interloper, which light the southwestern corner of the Full Moon.

East of Byrgius A, look for a bright ring of white light in the otherwise dark Mare Vaporum. That's **Gassendi,** a walled plain 68 miles (110 km) in diameter. Several central peaks can be seen within, as can a cracked floor. Several other bright rings can be seen across the Moon on the nights around the Full phase, including Posidonus, near Mare Serenitatis, and Theophilus, near Mare Nectaris.

The crater Manilius, near Mare Vaporum, also puts on a dazzling display around Full Moon, looking like a searchlight to the northeast of the lunar disk's center. **Proclus,** another beacon seen near Mare Crisium, displays a bright ray system when the Sun is high in its sky. The crater itself only measures 17 miles (27 km) across, but its rays extend for more than 100 miles (160 km). Southward along the Moon's eastern edge, you will also see the bright but small craters **Langrenus M, Stevinus A,** and **Furnerius A.** All three are made prominent by their bright, surrounding ejecta fields, which also all but obliterate their larger namesake craters. Finally, the craters **Manilius** and **Menelaus** shine brightly just north of the lunar disk's center.

### Waning Gibbous: Days 15 through 19

After Full Moon, the terminator switches to the eastern limb and now marks the sunset line. Once again, this is the region where we see the highest degree of shadow relief and, for the most part, the highest degree of visual interest. With sunlight now coming from the opposite direction, various features take on a different perspective during these waning phases (Figure 2.6) than they may have during the waxing phases.

Two of the first major features that sunset encroaches upon are Mare Crisium and Mare Fecunditatis. Remember the twin craters Messier and Messier A and the unique twin rays spouting from the latter? The rays are now all but extinguished, allowing observers to see the craters themselves. Messier, to the east, is conspicuously oval, while Messier A reveals itself to be a double crater, with a younger crater almost perfectly overlapping an older impact site.

Several prominent rilles twist throughout the jumbled terrain that lies to the south of the crater **Manilius.** Among the most obvious are **Ariadaeus Rille** and the **Hyginus Rille.** Both measure some 135 miles (220 km) long and extend approximately east-west. Switch to high magnification after you have located them. If you look carefully, you may notice that the Hyginus Rille is formed partially by a string of several, very closely spaced craters. Be sure to catch a last glimpse of the craters **Aristoteles** and **Eudoxus,** and the **Alpine Valley,** as all are slowly swallowed up by the frigid dark of the lunar night. The southern highlands, highlighted by craters **Maurolycus, Tycho,** and **Clavius,** also take on a rugged magnificence as sunset nears. Many other prominent craters to the lunar west, including **Copernicus, Kepler,** and **Plato,** remain visible until Last Quarter.

### Last Quarter: Days 20 through 22

Now a week after Full Moon, Last Quarter Moon (Day 21; Figure 2.7) offers a perspective on the scene different from what we studied closely at First Quarter. Last Quarter rises around local midnight, give or take an hour or so (possibly more, especially in the summer with daylight saving time), but remains high in the southern sky after sunrise so it can easily be viewed in the morning.

**Tycho's** ray system has disappeared, so this crater now looks like just another scar in the Moon's heavily cratered southern highlands. **Clavius** again dominates the scene. Moving northward along the terminator, again take a look for **Arzachel, Alphonsus,** and **Ptolemaeus. Hipparchus** and **Albategnius** ride right along the terminator on the night of Last Quarter, the latter's central peak appearing like a firefly hovering within a bright ring of light that marks the crater's rim.

Perhaps the biggest change due to the different lighting angle is seen in the **Straight Wall.** Recall how on the night after First Quarter, the Straight Wall looked like a thin, black scar along the eastern shore of Mare Nubium. At that time, we were not seeing the wall itself, but rather its darkened face that was still unlit by the Sun as well as the cliff's shadow. With sunlight striking the wall directly, it now appears as a bright, white line and is made especially prominent by the darker surrounding mare. The difference is quite striking.

Last Quarter is also a wonderful phase for viewing the craters **Aristillus, Autolycus,** and **Archimedes.** All three can be found just north of the terminator's center. Aristillus and Autolycus see sunset on the night of Last Quarter,

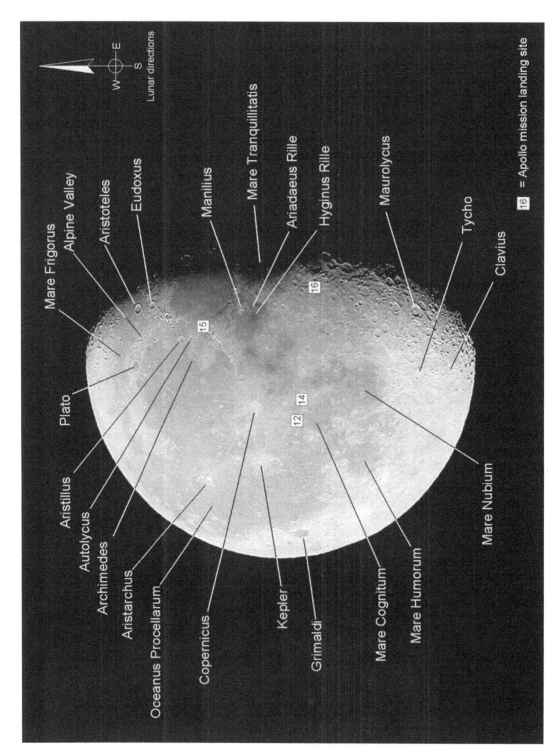

**Figure 2.6** *Eighteen-day-old Moon. Photograph taken by the author through an 8-inch f/7 reflector and a 26-mm eyepiece.*

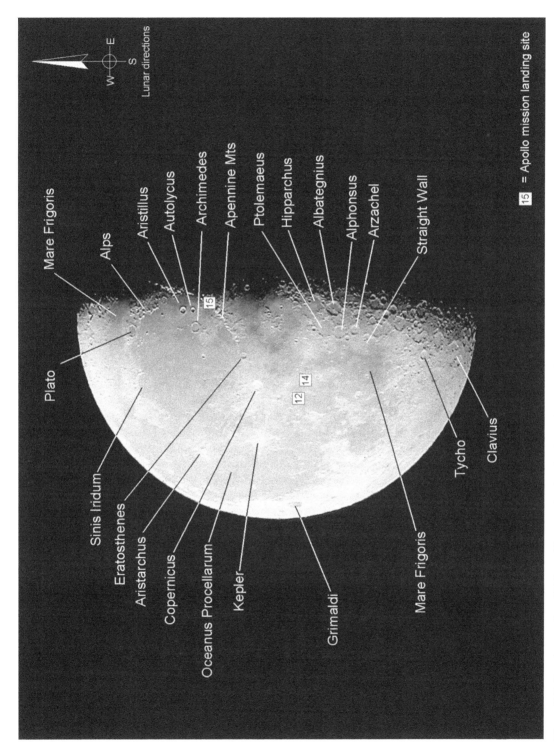

**Figure 2.7** *Twenty-one-day-old Moon. Photograph taken by the author through an 8-inch f/7 reflector and a 26-mm eyepiece.*

15 = Apollo mission landing site

while Archimedes remains sunlit for another 24 hours. Note the almost three-dimensional appearance of all three craters, as well as the ruggedness of the **Apennine Mountains,** to their south.

The lunar **Alps** also show up wonderfully in three-dimensional relief at Last Quarter. Trace their slopes to the northeast, toward the flooded crater **Plato.** Using at least 200×, take advantage of the Sun's low angle in Plato's sky and try to find the five craterlets that lie on the crater's floor. Wait for a moment of steady seeing to get a good, clear view. Any atmospheric turbulence may blur the image just enough to veil the craterlets from view.

### Waning Crescent: Days 23 through 28

As the Moon approaches the end of a lunation, with the terminator slowly making its way toward the eastern lunar limb, the thin crescent (Figure 2.8) rises closer and closer to Earth's sunrise each night. Perhaps the most obvious feature at this phase is the thin crescent within the crescent. The western-facing slope of **Sinus Iridum,** seen near the northern limb, is now bathed in shadow, with sunlight just catching its arc as the surrounding surface of **Mare Imbrium** is lost to the night on Day 24.

Sunset over **Copernicus** on Day 23 is quite striking. If time permits, watch over a period of hours as the lunar night slowly encroaches on the crater's western walls, sliding across the crater floor, then up the central peak and, lastly, over the eastern rim. **Kepler** remains sunlit until Day 25. With its brilliant rays now greatly subdued, look for some fine detail in the crater's uneven floor as well as in the irregular surface of **Oceanus Procellarum.**

**Gassendi** is a prominent walled plain during the waning crescent phases. Lying on the north shore of **Mare Humorum** (the Sea of Moisture), Gassendi displays an amazingly varied interior, with numerous hills, central peaks, thin, sinuous rilles, and even a pair of tiny craters.

Move toward the southern horn of the crescent and see if you can find an unusually long crater that looks like a giant footprint. That's **Schiller,** one of the oddest craters on the Moon. Schiller measures 110 miles by 44 miles (177 km by 71 km) across, a difference that is accentuated even more by foreshortening caused by the curvature of the Moon. It must have taken very unusual circumstances to create craters with this oval and oblong shape. To this day, the origin of Schiller remains a mystery. The only thing scientists agree on is that it almost certainly did not result from a single impact.

People who want to see the largest crater on the Moon are often disappointed to hear that it is actually quite difficult to spot. That's because **Bailly,** which lies so close to the southern lunar limb, appears very long but very thin due to foreshortening, going completely unnoticed unless the area is scanned carefully. To find Bailly, first zero in on Schiller. With a magnification of 150× to 200×, look between it and the edge of the Moon, past several small craters, for Bailly's shallow rim. Bailly measures 188 miles (300 km) across, nearly 30 miles (48 km) larger than Clavius, which is so much more conspicuous when in view.

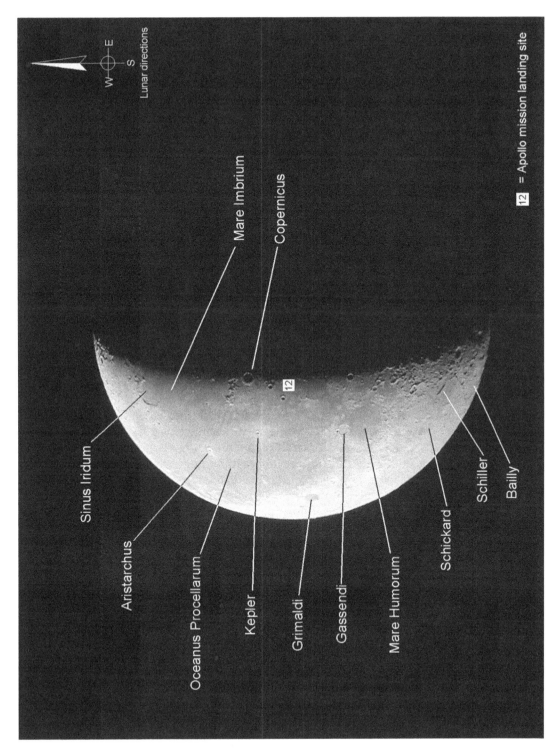

**Figure 2.8** *Twenty-four-day-old Moon. Photograph taken by the author through an 8-inch f/7 reflector and a 26-mm eyepiece.*

Though far smaller than Bailly, **Schickard** is much more prominent on the nights leading toward New Moon. You will find it just to the west of Schiller. Study Schickard closely, especially its unusual floor, which shows two darker regions sandwiching a brighter central plateau. Lunar geologists are not sure exactly what caused this apparent sinking of part of Schickard, but the effect is especially notable under the right lighting conditions.

Lastly, we come to the flooded impact basin **Grimaldi.** First highlighted at Full Moon, Grimaldi shows a perfectly smooth surface save for a small craterlet near its northwestern edge. Two sinuous rilles lie to the west and northeast of Grimaldi, although both prove challenging to spot due to the stark lighting.

### Lunar Occultations

Over the course of every lunar cycle, the Moon will pass in front of, or *occult*, many stars. Slowly, silently, the Moon creeps ever closer to a distant star, which appears to hover as the two draw closer. The star will look like a distant fire on the lunar horizon until, suddenly, it's gone. Extinguished without a trace. But as the Moon continues its patrol around Earth, the star will reappear in under an hour, since that's how long it takes the Moon to move eastward the span of its own diameter.

Viewing these occultations (Figure 2.9) can be enjoyable and sometimes even exciting. The best phases for occultation watching are the crescents, either waxing or waning. The dimmer Moon makes it possible to see faint stars that are nearby. But another plus is that, thanks to earthshine, you can see the entire disk. This makes it far easier to estimate when and where the occulted star will disappear and reappear, something not possible beyond the quarter phases, since the unlit portion of the lunar disk is not easily visible.

Depending on your location, the Moon may appear to only skim past a star in what is called a *grazing occultation*. Rather than cloak the star completely, the Moon's rugged limb, or edge, with all of its high mountains and deep valleys, alternately covers and uncovers the star, causing it to blink on and off several times. Teams of amateur astronomers often get together and space themselves evenly along a north-south line to record the exact timing of a star's blinking. Later, the collected times can be used to draw a precise profile of the lunar limb.

It's especially interesting to watch the Moon occult a bright star, binary star, or star cluster. Some favorites that occasionally find their way behind the Moon include the bright stars Regulus in Leo, Spica in Virgo, and Antares (itself a binary star) in Scorpius; the star Aldebaran and the Pleiades and Hyades star clusters in Taurus; and the Beehive Cluster (M44) in Cancer. Binary stars are especially interesting, as they don't blink out all at once. Instead they fade away in two (or more) steps as each of the stars in the system passes behind the Moon. Even if you cannot resolve the individual stars, this step-down effect, from bright to dim to gone, can be quite apparent.

On comparatively rare occasions, the Moon will pass in front of a planet or asteroid. Watching the Moon occult a planet can be very interesting. The Moon is always interesting to view through a telescope, but watching the disk

**Figure 2.9** *Occultation of the planet Venus by the crescent Moon. Photograph by Richard Sanderson taken through the 20-inch f/10 Schmidt-Cassegrain at the Springfield (Mass.) Science Museum's observatory.*

of a planet appearing to float just above the lunar surface is a spectacular sight. Rather than blinking out suddenly like a star, a planet will be seen to slowly fade away over a period of a few seconds as the Moon's limb gradually blocks the planet's disk.

Higher magnification is best for watching an occultation, as that will let you isolate the part of the lunar disk that will be covering the object. Reappearance is a little more difficult, since it is difficult to tell exactly where the star will pop back into view. Try to estimate the spot as best you can, but use a moderate power to play it safe, at least until you get a feel for it.

### *Lunar Occultation Information*

Lunar occultation predictions are computed and published annually by the U.S. Naval Observatory in Washington, D.C. Predictions for the occultations of brighter stars are published in the Royal Astronomical Society of Canada's annual *Observer's Handbook*, as well as in both *Sky & Telescope* and *Astronomy* magazines. Information about upcoming occultations is also available on-line from the International Occultation Timing Association (IOTA) www.lunar-occultations.com/iota, or write to 2760 SW Jewell Avenue, Topeka, KS 66611.

### *Lunar Eclipses*

Two, three, or sometimes as often as four times a year the Full Moon will pass through Earth's shadow, creating a lunar eclipse (Figure 2.10). While a solar eclipse (detailed in chapter 4) is seen over only a small portion of the day side of Earth and changes in appearance depending on where an observer is located, a lunar eclipse can be viewed from across the entire night side of Earth (weather permitting, of course) and will appear exactly the same for all observers on Earth's night hemisphere.

Lunar eclipses come in three varieties, depending on how deeply the Moon intrudes into Earth's shadow. A *total lunar eclipse* takes place when the Moon's entire disk is enveloped in the central portion of the shadow, called the *umbra*. No two total lunar eclipses are exactly the same. Since Earth is surrounded by an atmosphere, our planet's shadow is never completely opaque. Instead, a small amount of sunlight is bent, or refracted, through the atmosphere and into Earth's shadow. Due to the refractive properties of our atmosphere, light from the blue end of the visible spectrum is scattered, while light from the red end more readily passes through. This results in a reddish cast to Earth's shadow (and is also why we enjoy red sunsets and sunrises). Sometimes, the umbra paints a bright red-orange hue onto the lunar surface, while other eclipses appear coppery-red or even brownish gray. A total eclipse's

**Figure 2.10** *A lunar eclipse is one of the night sky's most fascinating events to watch. This photograph of a partial lunar eclipse was taken by Brian Kennedy through an 8-inch f/10 Schmidt-Cassegrain telescope.*

color depends on the clarity of our planet's upper atmosphere. In general, a vivid eclipse will be seen when the air is free of particulate matter, while a darker total phase will be witnessed when the atmosphere is polluted with foreign particles, such as volcanic emissions.

We see a *partial lunar eclipse* when the Moon is positioned in such a way that only a portion of it dips into Earth's shadow. Depending on how deeply the Moon enters Earth's shadow, the eclipsed portion of the lunar surface may appear a dark-red or rust color, or simply a charcoal gray, because of the sharp contrast in brightness between it and the brilliant part of the Moon that remains outside the umbra.

*Penumbral lunar eclipses* occur when the Moon passes through only the faint outer portion of Earth's shadow, called the *penumbra*. None of the lunar surface is completely shaded by Earth's umbra; instead, observers see only the slightest dimming near the lunar limb closest to the umbra. Unless at least three-quarters of the Moon enters the penumbra, the eclipse may prove undetectable.

Although any telescope will show a lunar eclipse well, I recommend that you use a small telescope for the best view. Large apertures tend to gather so much moonlight that the view can prove uncomfortably bright. And you certainly do not want to use a Moon filter here, since that will dilute the umbra's subtle colors. I also recommend using a low enough magnification so that the entire lunar disk fits into view.

In addition to watching the eclipse unfold, keep an eye out for any stars that might be occulted by the eclipsed Moon. The brilliance of the Full Moon usually makes it impossible to see occultations, but not so during a lunar eclipse (especially during the total phase). In fact, the crimson hue of the Moon often heightens the color contrast between it and the occulted star, especially if the star is relatively bright. If you are viewing occultations during a lunar eclipse, follow the advice about magnification given earlier in the "Lunar Occultations" section of this chapter.

Table 2.1 on the next page lists upcoming lunar eclipses. I have chosen to list only umbral eclipses, since penumbral eclipses are comparatively unimpressive. For additional information, please check my book *Eclipse!* (John Wiley & Sons, 1997).

*Table 2.1* **Upcoming Lunar Eclipses: 2003–2015**

| Date | Type | First Contact (UT[1]) | Mid-Eclipse (UT[1]) | Last Contact (UT[1]) | Area of Visibility |
|------|------|------|------|------|------|
| 2003 Nov 9 | Total | 23:33 | 01:19 | 03:05 | North and South America, Europe, Africa |
| 2004 May 4 | Total | 18:49 | 20:31 | 22:13 | Asia, Africa, Australia, New Zealand |
| 2004 Oct 28 | Total | 01:15 | 03:04 | 04:54 | North and South America, western Europe, western Africa |
| 2005 Oct 17 | Partial | 11:34 | 12:04 | 12:33 | Northwestern North America, Australia, New Zealand, Asia |
| 2006 Sep 7 | Partial | 18:06 | 18:52 | 19:38 | Europe, Africa, Australia, New Zealand, central and eastern Asia |
| 2007 Mar 3–4 | Total | 21:30 | 23:21 | 01:12 | Eastern North and South America, Europe, Africa, western Asia |
| 2007 Aug 28 | Total | 08:51 | 10:38 | 12:24 | North and South America, Australia, New Zealand, Pacific, eastern Asia |
| 2008 Feb 21 | Total | 01:43 | 03:26 | 05:09 | North and South America, Europe, Africa |
| 2008 Aug 16 | Partial | 19:36 | 21:10 | 22:45 | South America, Europe, Asia, Africa, Australia, New Zealand |
| 2009 Dec 31 | Partial | 18:53 | 19:23 | 19:54 | Europe, Asia, Africa, Australia, New Zealand |
| 2010 Jun 26 | Partial | 10:17 | 11:39 | 13:01 | Pacific, western Alaska, Australia, New Zealand, eastern Asia |
| 2010 Dec 21 | Total | 06:33 | 08:17 | 10:02 | North and South America, Europe, western Africa, eastern Asia |
| 2011 Jun 15 | Total | 18:22 | 20:12 | 22:02 | South America, Europe, Africa, Asia, Australia |
| 2011 Dec 10 | Total | 12:45 | 14:31 | 16:18 | Europe, eastern Africa, Asia, Australia, New Zealand, Pacific, North America |
| 2012 Jun 4 | Partial | 09:59 | 11:03 | 12:07 | Asia, Australia, New Zealand, Pacific, North and South America |
| 2013 Apr 25 | Partial | 19:51 | 20:07 | 20:23 | Europe, Africa, Asia, Australia, New Zealand |
| 2014 Apr 15 | Total | 05:57 | 07:45 | 09:33 | Australia, New Zealand, Pacific, North and South America |
| 2014 Oct 8 | Total | 09:14 | 10:54 | 12:34 | Asia, Australia, New Zealand, Pacific, North and South America |
| 2015 Apr 4 | Total | 10:15 | 12:00 | 13:45 | Asia, Australia, New Zealand, Pacific, North and South America |
| 2015 Sep 28 | Total | 01:06 | 02:47 | 04:27 | Eastern Pacific Ocean, North and South America, Europe, Africa, western Asia |

1. UT stands for Universal Time, the same as Greenwich Mean Time (GMT), the time zone referenced by astronomers to standardize their observations.

# 3

# The Planets and Asteroids

Some of the most fascinating views through a telescope or binoculars are of our planet's immediate family, the other planets in our solar system. All have something to offer, whether it's exquisite details in their thick, cloudy atmospheres, subtle detail on their barren surfaces, or simply the thrill of the hunt. Perhaps best of all, the brighter planets—Venus, Mars, Jupiter, and Saturn—transcend light pollution. They are bright enough to be seen even from the downtown area of any major city and so can be enjoyed by urban astronomers as readily as stargazers out in the countryside.

In terms of details visible, Jupiter is everyone's favorite, with its beautiful, belted atmosphere and the dance of its many moons. Saturn is no less wonderful, thanks to the complex ring system girding the planet's equator. Whenever I show Saturn to my students or members of the public, their immediate reaction is "it looks just like a picture." Mars, however, which everyone expects will show a finely detailed surface, often proves disappointing at first. Except when the Red Planet passes closest to Earth, its small size and great distance combine to produce a small, ill-defined world. But as your eye becomes more adept at sky watching, you will be amazed at how much Martian detail you can actually see, even when conditions are not their best. Detail on Venus, which shines so brilliantly in the sky, is hidden by the planet's opaque atmosphere. Still it is a beautiful world to visit as we watch it pass through phases like the Moon. Mercury, Uranus, Neptune, and Pluto, as well as most comets and hundreds of asteroids, are all difficult to spot, but that may well add to their appeal. It takes a special dedication to spot these worlds, which often go unseen by many stargazers.

Our tour of the solar system begins with Mercury and works its way outward. Mercury, Venus, Earth, and Mars are called the *terrestrial planets* because all of them are made primarily of rock and metals. Other traits that the terrestrial planets have in common include thin atmospheres, few (if any) moons, comparatively slow rotation on their axes, and close proximity to the Sun.

The four *giant*, or *Jovian* planets—Jupiter, Saturn, Uranus, and Neptune—have very different characteristics. These huge globes are comprised primarily of hydrogen and helium, with fractional percentages of other constituents. Many people erroneously call these the "gas giants," but in reality, very little of their hydrogen and helium is in a gaseous state. Instead, their inner pressures are so great that both elements are liquefied. Other qualities shared by the four giant planets include their great distance from the Sun, their rapid rotation (all rotate in fewer than 11 hours), and their large diameters. All four are also encircled by many natural satellites as well as ring systems.

Pluto, currently the farthest planet from the Sun, is not a member of either planetary group. Like the Jovian planets, Pluto is very far from the Sun. But because of its small diameter, only 1,400 miles (2,300 km) across, it is more like a terrestrial planet; in fact, it is smaller than any of the terrestrials. That has raised the issue in the past of whether or not Pluto is really a planet in the true sense of the word. Happily, that argument was finally put to bed a few years ago with the answer that yes, it is a bona fide planet. Having said that, however, Pluto's physical attributes, including its small size and frozen surface, are much more reminiscent of some of the Jovian planets' satellites than of one of the planets.

You'll also find a discussion about observing asteroids in this chapter as well as a section on how to observe comets. As the Greek root of their name implies, asteroids appear starlike through Earth-based telescopes. An asteroid may only look like a point of light, but tracking its slow progress against the background of stationary stars can be great fun. Comets are more readily distinguishable than asteroids because of their fuzzy appearance. While most comets are very faint, and therefore difficult to spot through amateur telescopes, a bright one will come along every now and then that catches everyone's attention.

## Mercury

FINDING FACTOR:  **
"WOW!" FACTOR:     Binoculars: *      Small telescopes (3″ to 5″): *
                  Medium telescopes (6″ to 8″): *

Despite the fact that it is one of the closest planets to Earth, third only to Venus and Mars, Mercury is one of the least observed and least understood members of our solar system. As they say in real estate, location is everything. That is especially true with Mercury. The closest planet to the Sun at 36 million miles (57.9 million km), Mercury is forever trapped in either the morning or evening twilight sky as seen from Earth. Often the challenge in seeing Mercury is simply in finding it!

Professional astronomers are not much better off than amateurs when it comes to viewing the innermost planet. Mercury remains so close to the Sun that the Hubble Space Telescope cannot chance pointing its way because of the danger that the nearby, intense sunlight poses to the telescope's sensitive instruments. Mercury holds the dubious distinction of being the only planet never to have been viewed by the Hubble.

Indeed, this lonely planet has gone completely ignored by the world's space programs, save for a single spacecraft, *Mariner 10*, which flew by Mercury three times in 1974. Because of Mercury's slow rotational rate of only 59 days, *Mariner 10* photographed less than half of Mercury's arid surface. Mariner revealed that airless, waterless Mercury is covered with craters and other impact features, including the Caloris Basin, measuring some 800 miles (1,300 km) across, 27 percent of the planet's 3,030-mile (4,879-km) diameter! The fact that Mercury wasn't shattered by the collision that created this feature is amazing in itself.

It should come as no surprise that, because of its proximity to the Sun, Mercury is a hellish world. Daytime temperatures top 800°F (425°C) at high noon! But since Mercury has no appreciable atmosphere to trap that heat during its long night, the temperature drops to −280°F (−173°C). This effect, called *radiant cooling*, causes Mercury to suffer the greatest temperature swings of any planet in the solar system.

### When and Where to Look

Although Mercury can become quite bright in the sky (magnitude −1 at its brightest to +1 at its dimmest), it never gets farther than 28° from the Sun, forever trapping it in the glow of twilight. Since Mercury sets no more than an hour after the Sun and rises no more than an hour before, you need a near-perfect horizon to stand any chance of seeing the planet. I can recall spotting Mercury often in the evenings some twenty years ago as I stood aboard the Staten Island Ferry, crossing New York Harbor. The view was especially memorable as the ferry passed the Statue of Liberty, although trying to determine which was Mercury and which were low-flying aircraft on approach to Newark International Airport could be tough at times.

The best chance for seeing Mercury is when the planet is either at *greatest eastern elongation* or *greatest western elongation*. Mercury is said to be at greatest eastern elongation when it is at maximum distance east of the Sun, which places it just above the western horizon after sunset. Greatest western elongation occurs when Mercury is at its farthest distance west of the Sun, making it visible in the east just before sunrise.

To really get the best view of Mercury, observers in the Northern Hemisphere should look for the planet when it reaches greatest eastern elongation in the late spring or greatest western elongation in late autumn. During those seasons, the *ecliptic* (the path in the sky followed by the Sun, Moon, and planets) is tilted at the steepest angle to the horizon. Take a look at the two drawings shown in Figure 3.1. Both show Mercury as it appears in the western sky at greatest eastern elongation. In each case, Mercury is 28° away from the Sun. Panel A, depicting the late spring, shows the ecliptic tilted nearly perpendicular to the horizon, maximizing Mercury's elevation. However, Mercury

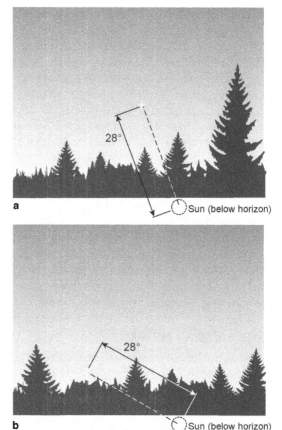

**Figure 3.1** *The angle that the ecliptic makes with respect to the horizon depends on the season. These views, as seen from the Northern Hemisphere, show how Mercury when at eastern elongation can appear significantly higher in the spring evening sky (a) than in the autumn (b), even though both it and the Sun are separated by the same angular distance.*

would be tough to see when situated as it is in panel B. Why? Because even though it is still 28° from the Sun, it is only a fraction of that distance above the horizon. In this case, the ecliptic is nearly parallel to the horizon, keeping Mercury low in the evening sky, as it is late in the autumn.

To spot Mercury, first check predictions to see when it is supposed to be visible in the sky, if at all. Table 3.1 lists the best times to Mercury-watch between 2003 and 2011.

### What You'll See

**With the Naked Eye and through Binoculars.** Although Mercury is bright enough to be seen with the eye alone in the evening and morning skies when it is at or near greatest elongation, binoculars may still be needed to find Mercury if it is hidden amid the haze and clouds that often accompany the twilight glow. Slowly scan back and forth just above the horizon where Mercury is expected to be, looking for a bright, flickering point of light immersed in deep morning or evening twilight. Remember, atmospheric conditions must be quite clear to be successful. Only attempt to spot Mercury when there is little or no smoke

Table 3.1 **Mercury Elongations: 2003–2011**

| Date | Elongation | Date | Elongation | Date | Elongation |
|------|-----------|------|-----------|------|-----------|
| **2003** | | **2004** | | **2005** | |
| Feb 3 | W | Jan 17 | W | Mar 12 | E |
| Apr 16 | E | Mar 29 | E | Apr 26 | W |
| Jun 2 | W | May 14 | W | Jul 8 | E |
| Aug 14 | E | Jul 26 | E | Aug 23 | W |
| Sep 26 | W | Sep 9 | W | Nov 3 | E |
| Dec 8 | E | Nov 20 | E | Dec 12 | W |
| | | Dec 29 | W | | |
| | | | | | |
| **2006** | | **2007** | | **2008** | |
| Feb 23 | E | Feb 7 | E | Jan 21 | E |
| Apr 8 | W | Mar 21 | W | Mar 3 | W |
| Jun 20 | E | Jun 2 | E | May 13 | E |
| Aug 6 | W | Jul 21 | W | Jul 1 | W |
| Oct 16 | E | Sep 29 | E | Sep 10 | E |
| Nov 25 | W | Nov 8 | W | Oct 22 | W |
| | | | | | |
| **2009** | | **2010** | | **2011** | |
| Jan 4 | E | Jan 26 | W | Jan 9 | W |
| Feb 13 | W | Apr 8 | E | Mar 22 | E |
| Apr 26 | E | May 25 | W | May 7 | W |
| Jun 13 | W | Aug 6 | E | Jul 19 | E |
| Aug 24 | E | Sep 19 | W | Sep 2 | W |
| Oct 5 | W | Dec 1 | E | Nov 14 | E |
| Dec 18 | E | | | Dec 22 | W |

*At greatest eastern elongations (listed as "E"), Mercury is situated to the east of the Sun and will be visible in the evening, low in the western sky right after sunset. Greatest western elongation ("W") places Mercury west of the Sun, visible in the eastern sky right before sunrise. Not all elongations are as favorable as others. In general, western elongations are most favorable in the autumn from the Northern Hemisphere and in the spring from the Southern. The opposite is true for eastern elongations.*

or fog or no clouds hugging the horizon, since even the slightest interference could render Mercury undetectable.

**Through a Telescope.** If you think spotting Mercury is tough, just try to see any surface detail on its scorched surface. About the best that most amateur telescopes can reveal is that—since it circles the Sun inside Earth's orbit—Mercury goes through phases similar to our own Moon and Venus. There are times when Mercury will appear as a thin crescent and other times as a half disk. A 3-inch (7.5-cm) telescope is large enough to reveal these phases, but only if our atmosphere is very calm and steady. Focusing your telescope is the tough part, however, since seeing conditions are often very poor at such a low altitude in the sky. Because of turbulence in our atmosphere, Mercury will often appear as little more than a tiny, unfocusable disk of light.

As speedy Mercury races around our star in its 88-day orbit, it alternately passes in front of and behind the Sun as seen from Earth. These points in its

Table 3.2 **Upcoming Transits of Mercury: 2003–2019**

| Date | Area of Visibility |
| --- | --- |
| 2006 Nov 8 | North and South America, Australia, Pacific |
| 2016 May 9 | North and South America, western Africa |
| 2019 Nov 11 | North and South America, western Africa |

orbit are called *inferior conjunction* and *superior conjunction,* respectively. Although the tilt of Mercury's orbit usually takes it either above or below the solar disk at conjunction, occasionally all three bodies line up. When this occurs at inferior conjunction, Mercury is seen to cross, or *transit,* the Sun. In a sense, a transit of Mercury is a very tiny eclipse. Since Mercury's disk is so small, however, it appears only as a tiny dot slowly making its way across the Sun's face. In fact, it looks almost like a sunspot but with an important difference. This sunspot moves!

Transits of Mercury can be great fun to watch. All that is needed is a 2-inch (5-cm) telescope and about 75×, but I must stress the need for caution. As discussed in the next chapter, viewing the Sun can be harmful. You should never look directly at the Sun with a telescope, binoculars, or your unaided eyes. Therefore, a telescope must either be used to project the Sun's image or outfitted with a properly designed solar filter in order to guard against possible damage to your eyes. Fortunately, well-designed filters are quite easy to come by.

Table 3.2 lists all Mercury transits occurring to the year 2019. As the dates draw near, be sure to check one of the astronomical periodicals, like *Astronomy* or *Sky & Telescope* magazine, for times and details.

Notice how each of the Mercury transits occurs within a few days of either May 8 or November 10. Coincidence? Not really. From our vantage point, Mercury's orbit, which is tilted 7° with respect to Earth's, intersects the ecliptic at two points, or *nodes,* that cross the Sun each year on those dates. If Mercury happens to be passing through inferior conjunction around that time of year, a transit will be witnessed somewhere on Earth.

## Venus

FINDING FACTOR:  *

"WOW!" FACTOR:    Binoculars: *        Small telescopes (3″ to 5″):  **
                 Medium telescopes (6″ to 8″):  **

Although it is often called Earth's "sister planet," Venus actually parallels our world only in terms of diameter and mass. Earth measures 7,921 miles (12,756 km) across, and Venus is 7,517 miles (12,104 km), but the family resemblance ends there. For while Earth's life-supporting surface is covered with about 70 percent water and shrouded with an atmosphere of 78 percent nitrogen and 21 percent oxygen, Venus's impenetrable atmosphere is 96 percent poisonous carbon dioxide. Sunlight diffuses through the atmosphere, just as it does on Earth. The difference is that Earth's surface is able to radiate energy through our atmosphere and back into space. The dense, carbon-dioxide atmosphere surrounding Venus, however, traps and retains the heat radiating back from the surface of the planet. As a result, the temperature at the Venusian surface is a constant 900°F (480°C), the hottest surface temperature of any planet in our solar system and even hotter than the self-cleaning setting on most ovens! This condition is known as the "runaway greenhouse effect" and is similar to what environmentalists are afraid could happen to Earth (but on a far more conservative scale) if pollution from combustion engines continues unabated. If this weren't bad enough, there is the issue of rain on Venus. Not water rain like here on Earth, mind you, but raindrops of sulfuric acid. A thick cloud layer of sulfuric acid in the planet's upper troposphere exists because sulfur dioxide from the planet's many volcanoes combines with what water vapor there is in the atmosphere. Just as water vapor condenses and falls to Earth in the form of rain, so too does the sulfuric acid in Venus's clouds condense and pelt the planet's arid surface.

Venus's atmosphere is completely opaque, and it is impossible for a telescope to penetrate it directly in order to see what lies below. Instead, Earth-based telescopes show only a disk covered in a solid deck of white clouds. Even the eye of an orbiting spacecraft cannot see through the clouds, although special ultraviolet filters do let planetary astronomers view and study the circulation of the planet's clouds.

### When and Where to Look

Like Mercury, Venus is only visible in the west after sunset or in the east before sunrise, never appearing more than 47° from the Sun. Appendix B lists the location of Venus (as well as Mars, Jupiter, and Saturn) in our sky from 2003 to 2015.

### What You'll See

**With the Naked Eye and through Binoculars.** To many, Venus will forever be known as the "Evening Star" when it is visible in the west and the "Morning Star" when visible in the eastern predawn sky, although, of course, it is not a star at all. There is no doubt, however, that Venus exhibits the brilliant jewel-like quality of a sparkling diamond against the fiery hues of twilight, far outshining all other stars and planets in the sky. Only the Sun and the Moon appear brighter. In fact, Venus can actually cast shadows when at greatest brilliance. At that point, it shines at better than magnitude −4.5 and may be glimpsed in broad daylight, if one knows exactly where to look.

**Through a Telescope.** From Earth, observers can watch as Venus goes from a thin crescent near inferior conjunction to a broad gibbous just before or after superior conjunction. Back in 1609, Galileo was the first person to record Venus's changing phases. His observations proved that the Sun was at the center of the solar system, as proposed by the Polish astronomer Nicholas Copernicus about a century earlier. Up until then, most people assumed that

### Attempts to View Venus

Eight of the unmanned *Venera* spacecraft sent by the then–Soviet Union from 1970 to 1981 completed the first, and thus far only, successful landings on Venus's surface. The harsh conditions, high temperature, and crushing atmospheric pressure caused each *Venera* craft to succumb two hours or less after touchdown (*Venera 14* set the survival record—127 minutes—in 1981), although some of the missions were able to return some fascinating photos of a very alien world.

More recently, the United States imaged the surface of Venus from an orbiting space-craft called *Magellan*. Rather than photograph the surface by conventional means, *Magellan* used high-resolution imaging radar to penetrate the clouds. The results revealed a world that is littered with impact craters, volcanic activity, and unusual tectonic forces. These same forces have created complex patterns of cracks and ridges, as well as high, continent-size plateaus. The largest, named Aphrodite Terra, is about the size of Africa. You can find more information on the Magellan mission to Venus on the NASA Jet Propulsion Laboratory's Web site, www.jpl.nasa.gov/magellan.

everything revolved around Earth in perfect circles. If Venus indeed orbited Earth, then Galileo would have also seen Venus go through a "full" phase. With Venus actually traveling around the Sun, full phase occurs at superior conjunction, when Venus is invisible. Were Venus orbiting Earth, it would also appear the same size regardless of phase. But as it is, Venus appears largest during the crescent phases and smallest when a gibbous, proving that its distance to us changes. Figure 3.2 shows why this is so.

Even the most modest telescope can easily repeat Galileo's historic findings. Venus measures about 1' of arc in diameter when closest to Earth, which unfortunately coincides with inferior conjunction. Although invisible at that point due to glare from the Sun, Venus appears only a little smaller for many weeks before and after.

As Venus pulls farther away from the Sun in our sky, the phases slowly progress from crescent to a half-illuminated disk, then to a gibbous shape. All the while, the apparent diameter of the planet is shrinking as the distance between our worlds grows. Higher magnifications are needed to resolve the phase clearly at this point, especially when Venus nears superior conjunction, its greatest distance from Earth. While Mercury moves through its series of phases in a matter of weeks, Venus takes several months to go from crescent to gibbous, and vice versa, since it takes the planet 224.7 days to orbit the Sun.

Although most people see Venus as a brilliantly white but starkly blank disk, some observers in the past have recorded vague dark markings on the planet. These were probably either an optical illusion or just imagination, since we know from extensive studies by several spacecraft that have visited our sister world that the carbon dioxide clouds blanketing Venus are impenetrable. One unwelcome feature you might see, depending on the type and quality of the telescope you are using, is a colorful halo of purple, blue, yellow, or green surrounding Venus. That is merely an optical flaw in the instrument called *chromatic aberration*, the result of a lens or lenses dispersing

Evening visibility                                          Morning visibility

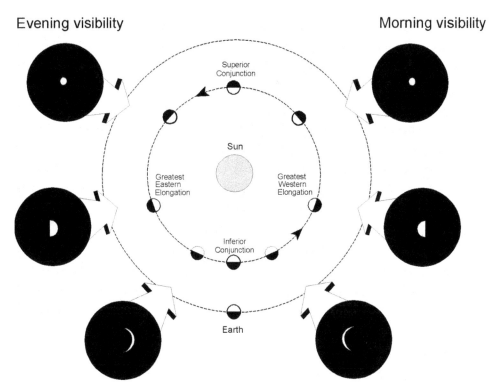

**Figure 3.2** *The planet Venus's orbit of the Sun as seen from Earth. Shown at each point in the planet's orbit is the appearance of the planet as seen through an Earth-based telescope.*

light. The effect is commonly seen in most inexpensive refractors, since chromatic aberration is an optical property of lenses. Although mirror-based reflectors do not inherently suffer from chromatic aberration, flares of false color can occur if a poorly designed eyepiece is used.

Just as Mercury occasionally transits across the Sun, so too does Venus. Transits of Venus are exceedingly rare events indeed. They occur in pairs separated by eight years, with over one hundred years between successive pairs, as you can tell from Table 3.3. The last transits of Venus were seen in 1874

Table 3.3 **Transits of Venus**

| Date | Area of Visibility | |
| --- | --- | --- |
| 2004 Jun 8 | Start-to-finish: | Central and western Asia, eastern and central Europe, eastern Africa, within the Arctic Circle |
| | Start only: | Eastern Asia, Australia, New Zealand |
| | Finish only: | Western Europe, western Africa, eastern North America |
| 2012 Jun 6 | Start-to-finish: | Pacific Ocean, New Zealand, within the Arctic Circle |
| | Start only: | North and South America |
| | Finish only: | Eastern Asia, Australia |

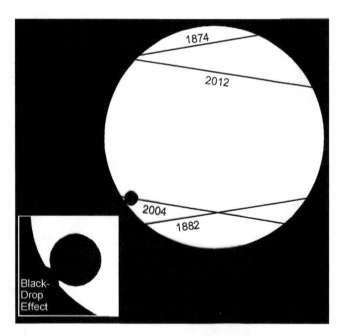

**Figure 3.3** *A sky event that no one alive has ever seen: a transit of the planet Venus across the Sun's disk. The next transits of Venus, in June 2004 and June 2012, are sure to attract wide attention from the astronomical world. The lines show Venus's projected tracks across the Sun's disk during both upcoming transits. Insert: The much-anticipated "black-drop effect" will be visible when Venus appears to pass over the edge of the solar disk at the beginning and end of each transit.*

and 1882, while the next pair will take place in 2004 and 2012. Talk about good timing! But again, just like watching a transit of Mercury, caution must be exercised when viewing one of the Venus transits as well. Never look directly at the Sun with a telescope, binoculars, or your unaided eyes. Instead, a telescope must either be used to project the Sun's image or outfitted with a properly designed solar filter in order to guard against possible damage to your eyes. The next chapter offers further advice on how to observe the Sun safely.

While no one alive at the time of this writing has ever seen a transit of Venus, we can envision what will be seen on those dates, based on records from the nineteenth century. Unlike Mercury, which appears as a perfectly round disk at the time of transit, Venus will look slightly diffuse because of light refracting through its atmosphere. Higher magnification telescopes will also display an unusual "black-drop" effect (Figure 3.3) at second and third contacts, when Venus appears to "pull" away from the Sun's edge, much as a raindrop drips away from a leaf. No one knows for sure what causes the black drop. Some say it is caused by diffraction or atmospheric refraction, but many authorities believe that neither fully explains the appearance.

## Mars

FINDING FACTOR:  * to **
"WOW!" FACTOR:    Binoculars: *      Small telescopes (3″ to 5″):  **
                  Medium telescopes (6″ to 8″):  **

No planet has piqued our curiosity as much as the Red Planet, Mars. First worshiped as the god of war, later revered or feared as the home of bizarre, alien life, Mars is now seen as a planet to be studied, visited, and perhaps ultimately colonized. What is it about the Red Planet that grabs mankind's attention?

Interest in Mars was kindled long before the invention of the telescope, but it seemed to take on a life of its own after a fateful telescopic observation by Italian astronomer Giovanni Schiaparelli in 1877. That year Schiaparelli announced that he saw *canali*, or "channels," on Mars. Overzealous English-speaking astronomers misinterpreted this to mean "canals," artificial waterways constructed by intelligent beings. Suddenly, Mars was inhabited!

Among the leading advocates of the idea of an advanced civilization on Mars was the American astronomer Percival Lowell. He and others quickly accumulated a staggering number of drawings of "canals" crisscrossing the Red Planet. Many astronomers did not see or believe in Lowell's canals, but the public, and Hollywood, became captivated by the idea.

The speculation about highly advanced Martians was forever put to rest in 1964, when *Mariner IV* returned photographs of barren, heavily cratered Mars. But as visions of little green men faded, planetary enthusiasts grew more excited at the prospect of getting to know our neighboring world more intimately. The unmanned *Viking 1* and *Viking 2* spacecraft culminated the decade-long effort of the United States's Mars program when they soft-landed in the summer of 1976. And now, after more than twenty years, interest in Mars has been rejuvenated thanks to the successful Mars Pathfinder and Global Surveyor missions as well as the unblinking eye of the Hubble Space Telescope. Unfortunately, NASA suffered setbacks with the losses of the *Mars Climate Observer* and *Polar Lander* spacecraft in 1999 but was vindicated in late 2001 when the *Mars Odyssey* craft successfully entered orbit around the Red Planet. As this book is written, *Mars Odyssey* is returning some amazing photographs. Mission updates can be found on the Internet at mars.jpl.nasa.gov/odyssey.

From the *Viking 1* and *Viking 2* robotic landers in the 1970s as well as *Mars Pathfinder* in the 1990s we now know that Mars has an unusual composition and climate. The planet's reddish color is due to the soil's composition of mostly iron oxides and clays. In other words, Mars is rusty! And while the average temperature on Mars is about −67°F (−55°C), readings can fluctuate from as low as −207°F (−133°C) during the winter near the Martian poles to almost 80°F (27°C) on summer days at the Martian equator. Part of that wide range is due to Mars's atmosphere. A thick atmosphere, such as here on Earth, insulates a planet's surface from dramatic cooling at night. But radiational cooling due to the thin, carbon dioxide atmosphere that surrounds Mars lets any heat that builds up during the day escape back into space at night.

Mars is a desertlike world that has heavily cratered highlands in its southern hemisphere and younger volcanic plains over much of its northern hemisphere. The volcanic field known as the Tharsis Bulge is as large as North America and includes Olympus Mons, the largest volcano found in our solar system. The summit of Olympus Mons towers more than 15 miles (24 km) above the surrounding plain and spans more than 300 miles (485 km) in diameter at its base. Put another way, if the center of Olympus Mons were placed over New York City, the outer base would extend to Washington, D.C., Boston, and Montreal, Canada.

### When and Where to Look

Amateur astronomers are drawn to the Red Planet when Mars is well placed for observation, which only occurs once every twenty-six months, when our two planets are closest. This point, when Mars lies directly opposite the Sun in our sky and Earth-Mars distance is at a minimum, is referred to as *opposition*. Table 3.4 lists upcoming Martian oppositions.

As the gap between that distant world and ours closes, the Martian globe grows in apparent size to reveal a complex network of dim, shadowy surface features. Some oppositions of Mars are better than others. An *aphelic* opposition occurs at or near Martian aphelion, its farthest point from the Sun, when the Red Planet comes no closer than 50 to 61 million miles (81 to 98 million km) to Earth. During these comparatively poor viewing periods, Mars, which measures 4,219 miles (6,794 km) in diameter, will grow to only about 14 seconds of arc (abbreviated 14″) across. In more favorable years, when Mars reaches opposition at or near perihelion, when it is closest to the Sun, the planet will be less than 35 million miles (56 million km) from Earth and measure about 25 arc-seconds (25″) across. These are called *perihelic* oppositions. The Martian oppositions in 1995, 1997, and 1999 were aphelic, but those in 2003 and 2005 are perihelic, affording observers some prime Mars-watching. Table 3.4 shows how aphelic oppositions occur north of the celestial equator (indicated by the positive declination values in the rightmost column), while perihelic oppositions occur to its south (negative declination values).

### What You'll See

**With the Naked Eye and through Binoculars.** Let's begin our journey to the Red Planet by finding it. To the eye alone, Mars looks like a reddish orange star. Near opposition, especially when near perihelion, Mars looks like a brilliant ruby shining around magnitude −2.5, suspended against the starry backdrop. But at times other than opposition, Mars can be far tougher to pick out. As the distance to Mars increases, its brilliance drops rapidly, dwindling to nearly second magnitude, and its dazzling blood red color grows anemic.

*Table 3.4* **Oppositions of Mars: 2003–2014**

| Opposition Date | Diameter of Disk (seconds of arc) | Earth-Mars Distance (astronomical units) | Declination |
|---|---|---|---|
| 2003 Aug 28 | 25.1″ | 0.37 | −15° |
| 2005 Nov 7 | 19.8″ | 0.47 | +15° |
| 2007 Dec 28 | 15.5″ | 0.60 | +27° |
| 2010 Jan 29 | 14.0″ | 0.66 | +22° |
| 2012 Mar 3 | 14.0″ | 0.67 | +10° |
| 2014 Apr 8 | 15.1″ | 0.62 | −05° |

Like all planets, Mars orbits the Sun from east to west as seen from Earth. As a result, observers can follow Mars's march against the background stars. Because the Red Planet takes 686 days to orbit the Sun, while Earth only requires 365 days, our planet catches up to and laps Mars at every opposition. When this happens, we on the faster-moving Earth see Mars appear to slow up in the sky, stop its eastward motion, and actually reverse toward the west over a period of weeks. Called a *retrograde loop,* Mars will continue this apparent backward motion until the distance between the two planets increases enough to cancel the effect. Our ancient ancestors were baffled by this, thinking that Mars must be a god to have the power to move like this among the stars, but today we know that it's just because of the Red Planet's slow orbit.

Even during a perihelic opposition, Mars is just too small to reveal much of its true personality through binoculars. The increased magnification will certainly make plotting its motion against the stars more apparent, but apart from its distinctive red color, and perhaps just the slightest hint of the polar caps through 20× and higher power binoculars, Mars reveals little.

**Through a Telescope.** Even at its best, Mars only appears slightly larger than a ringless Saturn, and much smaller than either Venus or Jupiter. This means that we must be Mars-smart to see its hidden details. During a favorable opposition, the Martian polar caps, dark markings on the planet's surface, and even an occasional dust storm are all visible through a 3-inch (7.5-cm) or 4-inch (10-cm) telescope at about 100×. The key to success is to wait for favorable conditions.

Even when Mars is closest to Earth, it demands high-quality optics. In general, a refractor is the instrument of choice among planet watchers, followed by long-focus Newtonian and Cassegrain reflectors, since these usually produce the highest image contrast. Telescopes with large central obstructions caused by their secondary mirrors (e.g., catadioptrics and "fast" Newtonians) diminish image contrast, which degrades the view. Regardless of the telescope design, however, make certain that the optics are clean, well collimated, or aligned, and fully acclimated to the ambient outside temperature before attempting to view Mars.

Just as important as the telescope is the quality of your eyepieces, since magnifications in excess of 200× are usually needed to see fine details. Popular super-wide-field eyepieces offer wonderful, panoramic views of star fields and broad nebulae but are often surpassed by more conventional eyepieces, such as orthoscopics, Brandons, and Plössls, for planetary observing. Indeed, my favorite eyepiece for viewing Mars is an old 9-mm orthoscopic eyepiece that I purchased secondhand several years ago for $30.

Here's an eyepiece trick that's often used with great success by Mars observers. Instead of using a single, short-focal-length eyepiece, combine a longer-focal-length eyepiece with a high quality Barlow lens. The greater eye relief of most longer-focal-length eyepieces means that observers can view through them with less discomfort, reducing the eye-to-brain "noise" that can block fine details.

Veteran planetary observers also recommend using color filters to enhance various features on Mars. For instance, orange (Wratten #21) and red (#23A or #25) filters increase the contrast of the dark surface markings against the bright orange desert regions. Blue (#38A or #80A) and green (#58) filters are best for showing the polar caps as well as any haze, fog, or clouds in the planet's thin atmosphere. Keep in mind, however, that while they work well, color filters are not magic! They may help in revealing some latent details on the surface, but they are helpless at overcoming poor seeing conditions and inadequate optics.

The final element for successful Mars-watching is beyond our control: atmospheric conditions (on Earth). Mars will look its best when the air is calm and seeing conditions are steady. To judge this, simply look to the stars. Are they twinkling? If so, the poor seeing may well hamper your observations. But on nights when conditions are steady, Mars will have no choice but to reveal its secrets. The best nights for making planetary observations often look hazy to the eye. Clear air (that is, transparent air) is often accompanied by the passage of a weather front, where Earth's upper atmosphere can prove very turbulent. This makes spotting fine details on Mars nearly impossible. During the languishing nights of haze and humidity, however, our atmosphere is often very tranquil, affording good seeing.

Identifying specific Martian characteristics is one of the greatest challenges facing backyard planetary astronomers. Always remember the two *P*'s of Mars: patience and perseverance. There will be times when Mars will appear as little more than a quivering blob, but on that rare occasion when our atmosphere is tranquil, the *real* Mars will come through. Figure 3.4 serves as a quick guide to the best the Red Planet has to offer observers on Earth. Each of the four panels is centered on a different central meridian, or Martian longitude. For instance, the upper left panel is labeled "CM 0 degrees," which means that Martian longitude 0° passes vertically through that view. The other three panels show Mars at different central meridian settings, giving observers a complete view of the planet. Of course, the Martian central meridian will probably be some intermediate value when you view the planet, so you'll need to use a little imagination to adjust the view accordingly.

From Earth, the most prominent features of Mars are its two snow-white polar caps. Both the north and south polar caps appear to grow in size during the Martian winter and shrink during the Martian summer. Studies show that each Martian polar cap is actually two separate, though overlapping, features. The seasonal caps (that is, the portions that we see changing with the seasons) are made of relatively thin layers of frozen carbon dioxide (dry ice). The permanent polar caps are much thicker and do not change appreciably with the seasons. The south polar cap is also composed of frozen carbon dioxide along with some water ice, while the north cap is mostly water ice.

During an opposition, one polar cap is usually tilted toward Earth while the other is tilted away and therefore invisible. Aphelic oppositions find the northern polar cap tilted our way, while during perihelic oppositions, it's the southern cap's turn.

As we approach and pass opposition, it is interesting to watch for changes in whichever polar cap is visible, as it shrinks during that hemisphere's summer and grows during the Martian winter. Often with the thawing cap come

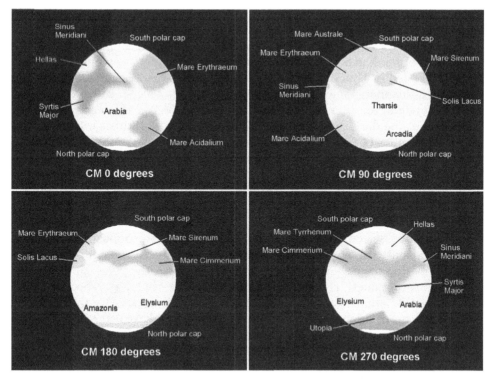

**Figure 3.4** *Four faces of Mars. Can you spot these elusive features through your telescope? Wait for the next opposition for your best chance. Each map shows the planet's surface centered on a different Martian latitude, called the Central Meridian (CM). Since Mars will probably not look exactly like one of these panels but rather will be at some intermediate point, you will need to estimate surface features accordingly.*

complex clouds high in the Martian atmosphere, which are best seen with a blue or violet filter.

Scattered across the red Martian surface are the famed dark markings. Research based on data received from spacecraft in Martian orbit over the years shows that these dark regions are thought to be underlying bedrock, while the bright surrounding areas are vast deserts covered with wind-blown dust. Although the Martian atmosphere is quite thin compared to Earth's, strong winds move sand and dust from region to region, often in spectacular dust storms. Over long periods of time, the large dark areas that we see through backyard telescopes remain pretty much unchanged, but smaller details visible through the Hubble Space Telescope as well as from orbiting spacecraft come and go as they are covered and then uncovered by sand and dust.

The most easily spotted dark feature on Mars is **Syrtis Major,** a triangular wedge extending north-south across the planet's equator. Many small, dark impact craters litter the area, although they cannot be distinguished by any telescopes here on Earth. Look for a circular feature just to the south of Syrtis Major. That's **Hellas,** now known to be a huge impact feature, the largest on Mars and the only Martian crater visible through amateur telescopes. **Sinus**

**Meridiani,** a claw-shaped patch of dark bedrock located along the Martian equator, is found about a fifth of the way around from Syrtis Major. The bright desert regions of **Amazonis, Arabia, Arcadia, Tharsis,** and **Elysium** dominate much of the Martian northern hemisphere. Planetary geologists are especially interested in Tharsis, as it is known to be the site of several major, albeit dormant, volcanoes, including **Olympus Mons,** the largest volcano found in the solar system. Another famous Martian feature, Valles Marineris, is located just south of the dark region **Mare Acidalium.** Although this series of valleys spans 3,000 miles (4,800 km), it remains invisible through amateur telescopes. Finally, be sure to keep an eye on Mars, as it may almost look like it's watching you! **Solis Lacus,** the "Eye of Mars," is a large, round dark area punctuated by a darker central region. Spacecraft have photographed the area partially covered in frost and fog, suggesting that there may be frozen water fairly near the surface.

Mars is best seen from about three months before opposition to about three months after. Watch it night after night throughout the course of a passage, comparing the view over several weeks. How does the polar cap compare? Is it larger or smaller? Have any clouds formed nearby? Pay especially close attention to the appearance of Syrtis Major and Solis Lacus. Are they growing darker and more apparent, or becoming smaller and weaker? Try different color filters and see how results compare. Normally, for instance, surface features on Mars appear rather vague and indistinct when viewed through a telescope equipped with a blue or violet filter. Occasionally, and without apparent cause, a "blue clearing" will occur that allows the Martian surface to be seen clearly for up to several days through these filters. Despite scrutiny of Mars by spacecraft and the Hubble Space Telescope, the cause of these blue clearings remains a mystery.

## Asteroids

FINDING FACTOR:   ** to ****
"WOW!" FACTOR:   Binoculars: **      Small telescopes (3″ to 5″):  **
                 Medium telescopes (6″ to 8″):  **

On January 1, 1801, Giuseppi Piazzi was observing the sky from his home in Sicily when he happened upon a faint starlike object in the constellation Taurus the Bull. No doubt he thought it strange that this star did not appear on his star charts of the region. He marked its position relative to the other, known stars and made it a point to return to that area on the next clear night. After a few days passed, he found, to both his surprise and delight, that his "star" had moved! Further observations proved that Piazzi's new object was orbiting the Sun between the orbits of Mars and Jupiter at a distance of 2.77 A.U. (astronomical units) away and taking 4.6 years to complete a circuit. Piazzi had found the first asteroid. As is customary, the discoverer of a new member of the solar system has the honor of naming it. Piazzi chose Ceres after the Roman goddess of Sicily. Ceres is often written as "(1) Ceres," the number indicating the order of discovery.

Since then, several hundred thousand asteroids have been discovered, and estimates suggest that there may be as many as 1 million that are 0.6 mile (1 km) across or larger. Each is assigned a sequential number based on its order of discovery, as well as a name. The first asteroids were named for Greek and Roman goddesses, although those were soon used up. Asteroids were then given other female names, but those too were eventually exhausted. Today, an astronomer discovering an asteroid often names it after a colleague as a way of honoring his or her contribution to the field of astronomy.

Where did the asteroids come from? While much of the material that formed the solar system congealed into the Sun or planets, some of the left-over material gathered into the vast expanse between Mars and Jupiter, eventually cooling to form the asteroids. While the planets have continued to evolve, the asteroids are thought to be unchanged since their formation 4.5 billion years ago.

Most asteroids are located between the orbits of Mars and Jupiter and fall into one of three major categories based on composition. *Carbonaceous* asteroids, meaning that they are rich in carbon, are the most common. The majority of these are very dark, reflecting only 3 percent to 4 percent of the sunlight that strikes them, although some carbonaceous asteroids reflect as much as 60 percent of the light reaching them. Other asteroids are stony in nature, possessing few of the dark carbon compounds seen in others. Finally, the third and rarest type of asteroids are metallic, composed of mostly iron and nickel.

### When and Where to Look

Looking for asteroids can be like hunting for a celestial needle in a haystack, but a good chart will help. Whenever an asteroid of about 8th magnitude or brighter is due to be visible, *Astronomy, Sky & Telescope,* and other periodicals will publish finder charts showing its path against the stars. Many software "planetarium" computer programs can also create finder charts that can be customized to your specific time, date, location, and telescope. Some of the more popular programs include The Sky, SkyMap Pro, Megastar, and Starry Night.

If you will be using binoculars to hunt for asteroids, be sure to support them on a tripod or other mount to make it easier to go back and forth with the chart without having to reaim them every time. With a telescope, start out with a low-power eyepiece. When you are certain that you are aimed at the correct star field, switch to a higher magnification to zero in on the asteroid. Look toward the area of sky where the asteroid is predicted to be. Once centered, accurately check the stars visible in the field against the asteroid's finder chart. Carefully note any geometric patterns that the stars form with the asteroid, such as a triangle, arc, or rectangle. Continue to play connect-the-dots until the asteroid eventually becomes evident. Make a drawing of the field, being careful to plot the locations of all stars in the field as accurately as possible. Then come back in a night or two, just like Piazzi, to find out if the "star" you believed was the asteroid has moved.

Table 3.5 **The Brightest Asteroids**

| Asteroid | Magnitude (maximum) | Diameter | |
|---|---|---|---|
| | | (miles) | (km) |
| (4) Vesta | 5.1 | 330 | 530 |
| (2) Pallas | 6.4 | 354 × 326 × 300 | 570 × 525 × 482 |
| (1) Ceres | 6.7 | 600 × 579 | 960 × 932 |
| (7) Iris | 6.7 | 124 | 200 |
| (433) Eros | 6.8 | 20 × 8 | 33 × 13 |
| (6) Hebe | 7.5 | 116 | 186 |
| (3) Juno | 7.5 | 149 | 240 |
| (18) Melpomene | 7.5 | <90 | <150 |
| (15) Eunomia | 7.9 | 159 | 256 |
| (8) Flora | 7.9 | 88 | 141 |
| (324) Bamberga | 8.0 | 143 | 230 |
| (1036) Ganymed | 8.1 | <90 | <150 |

### What You'll See

**Through Binoculars or a Telescope.** Asteroids, even when viewed through the largest telescopes on Earth, are seen as nothing more than starlike points of light. In fact, that is why they were given the name asteroid: *asteroid* means "starlike." The only way to identify an asteroid is to follow its movement against the more distant fixed stars. Over sixty asteroids become brighter than 10th magnitude and should therefore be visible through binoculars and small amateur telescopes. Table 3.5 lists the dozen brightest asteroids, along with their maximum magnitude and approximate size. Asteroids change very dramatically in brightness depending on their distance from Earth. The table quotes their maximum brightness, when their distance away from us is minimal. Notice that (4) Vesta will occasionally even crack the 6th-magnitude, naked-eye barrier. Interestingly, Vesta does not fit into any of the three categories mentioned above. Studies show that its surface is covered with basalts, which are volcanic in origin. Vesta is no longer volcanic, although at one time it must have been. How such a small body could be volcanically active remains a mystery.

## Jupiter

FINDING FACTOR:  *

"WOW!" FACTOR:    Binoculars: **    Small telescopes (3″ to 5″): ****
                 Medium telescopes (6″ to 8″): ****

Jupiter, the largest of all the planets, displays some amazing detail through just about any telescope. Even a quick glimpse will show that Jupiter is not round. Since gigantic Jupiter takes less than ten hours to rotate, the planet is actually squeezed by the momentum, which flattens it at the poles and causes it to bulge at the equator. Measurements show that Jupiter is 88,700 miles (142,980 km) in diameter at its equator but is 7 percent smaller from pole to pole.

Keep in mind that when we gaze Jupiter's way, we are not looking at the planet's surface per se but rather at the top level of an incredibly deep atmosphere. Studies show that its clouds are composed of about 90 percent hydrogen and 10 percent helium, with traces of methane, water, and ammonia. What lies beneath those clouds is anybody's guess, but current findings suggest that Jupiter has a core of solidified hydrogen rock and ice ten to fifteen times the mass of Earth. This core is surrounded by a sphere of liquefied metallic hydrogen.

Jupiter is emitting heat from deep within its core that is most likely left over from the planet's formation, but also possibly from Jupiter's slow contraction, which has continued since it was formed 4.5 billion years ago (even a minuscule rate of contraction can generate significant amounts of heat). The heat from Jupiter's core causes the atmosphere to circulate vertically, with lower layers rising to the top in the form of convection currents. These heated layers appear as lighter belts in Jupiter's turbulent atmosphere, while the darker belts are comparatively cooler, sinking back downward, only to be reheated and circulated back toward the visible surface again.

### When and Where to Look

Jupiter is the fourth brightest object in the sky after the Sun, Moon, and Venus, so spotting it is a simple task. To see if Jupiter will be in the sky tonight, check Appendix B, which lists the positions of the naked-eye planets through 2015.

### What You'll See

**With the Naked Eye and through Binoculars.** Although seeing detail on Jupiter itself requires a telescope's magnification, the dance of its larger satellites is a favorite sight among binocular observers (Figure 3.5). Sixteen moons orbit Jupiter, although only the brightest four are visible in binoculars and amateur telescopes. In 1610, Galileo became the first person to record their presence, and so today, we call them collectively the **Galilean satellites,** with each named for a mythological lover of the Roman god Jupiter. Innermost of the four is **Io,** which takes just 1 day 18 hours to complete an orbit. Next out, **Europa** circles Jupiter in 3 days 13 hours. **Ganymede,** the largest satellite in the solar system at 3,267 miles (5,262 km) in diameter, takes just under 7 days 4 hours to orbit, while the outermost Galilean satellite, **Callisto,** takes 16 days 16 hours.

All four Galilean moons orbit in the plane of Jupiter's equator, restricting their positions to directly east or west (left or right) of Jupiter's equator. Over several nights, they can be seen to change positions relative to one another and the planet itself. Sometimes, two will be on one side and two on the other; at others, there may be three on one side and one on the other. Perhaps only three, or even two, will be visible, with the others in eclipse behind Jupiter or transiting across the planet's face.

**Figure 3.5** *Jupiter and its moons (left to right) Ganymede, Io, Europa, and Callisto. Photograph by Chris Flynn through an 8-inch f/10 Schmidt-Cassegrain telescope.*

**Through a Telescope.** Jupiter's atmosphere puts on a spectacular show of alternating dark and light bands (Figure 3.6). The bright **zones** in Jupiter's atmosphere are composed of clouds of ammonia ice crystals, while the dark **belts** are regions of ammonium hydrosulfide and other compounds. Swirling eddies set into violent motion by rising heat currents and chemical reactions among the atmosphere's ingredients can be spotted in the maelstrom that marks where the belts and zones thrash together, testifying to the turbulent nature of Jupiter's atmosphere.

The most prominent features of Jupiter's clouds are the **North and South Equatorial Belts,** two parallel dark-brown belts bordering the white, central **Equatorial Zone.** All three features are visible through telescopes as small as 2 inches (5 cm) aperture, while 8-inch (20-cm) and larger instruments may further reveal that the South Equatorial Belt is actually divided in half by a pencil-thin bright zone.

Lying on the southern edge of the **South Equatorial Belt** is Jupiter's famous **Great Red Spot.** The Great Red Spot has been known since about 1665, when it was discovered independently by Giovanni Domenico Cassini and Robert Hooke (though there is some evidence that Hooke may have actually seen a different, now defunct spot in the North Equatorial Belt). The Great Red Spot certainly lives up to its name, for it measures some 15,500

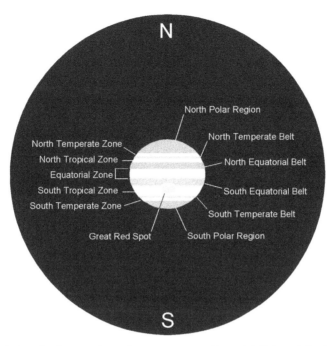

North Polar Region

North Temperate Zone
North Tropical Zone
Equatorial Zone
South Tropical Zone
South Temperate Zone

North Temperate Belt

North Equatorial Belt

South Equatorial Belt

South Temperate Belt

Great Red Spot

South Polar Region

**Figure 3.6** *Jupiter's striped atmosphere shows some magnificent detail through backyard telescopes. Try to identify as many of the striped bands as you can.*

miles long by 7,400 miles wide (25,000 km by 11,900 km). By contrast, Earth is only 7,921 miles (12,756 km) in diameter. Based on infrared observations conducted both by passing spacecraft and Earth-based telescopes, as well as the apparent direction of its rotation, astronomers believe that the Great Red Spot is a high-pressure region whose cloud tops are higher and colder than the surrounding regions. But why the Great Red Spot has persisted for more than 350 years remains a mystery.

While it measures up in greatness, the Great Red Spot is rarely red. Instead it tends to vary between an anemic orange, beige, salmon, and at times even white. Spotting the Great Red Spot requires a slow and methodical search; a quick glance just causes it to blend into the surroundings. Aperture is used to good advantage here, as the greater light-gathering ability of larger instruments will help accentuate the weak coloration. My 8-inch (20-cm) reflector rarely shows the Great Red Spot as more than a pale orange blemish. Of course, the Great Red Spot also might not be seen simply because it is on the far side of the planet at the time. If it is nowhere to be found at first, come back a few hours later and try your luck again.

Given good seeing conditions, higher magnifications will also reveal a multitude of other, less-prominent belts and zones across Jupiter. These include the **North** and **South Temperate Belts** and **Zones** and the **North** and **South Polar Regions.** Color filters can help highlight some of these subtle details. Try using a yellow-green (Wratten #11) or orange (#21) filter to pick out subtle features in the darker belts. A green (#56 or #58) or blue (#38A or #80A) filter helps accentuate the Great Red Spot's visibility. As unlikely as it sounds, I have also had success using a narrowband light-pollution filter, such as a Lumi-con UHC or Orion UltraBlock filter. Light-pollution filters are usually thought of as only useful on certain types of deep-sky objects, but the filtering properties of narrowband filters also seem to have a beneficial effect on Jovian features.

The amount of detail visible on Jupiter also depends greatly on the steadiness of the sky, the quality of your telescope's optics, and the planet's distance. Clearly, the closer we are, the larger the planet will appear and the greater the amount of detail seen. The best time to observe any of the outer planets, such as Mars, Jupiter, or Saturn, is when they are at or near opposition, the point in a planet's orbit when the planet lies directly opposite the Sun in our sky and its distance from Earth is at a minimum. Dates of upcoming oppositions can be found in astronomical periodicals, such as *Astronomy* and *Sky & Telescope.*

Through a telescope, you'll be able to follow the interplay of the Galilean satellites—Io, Europa, Ganymede, and Callisto—more closely than through binoculars. Try to figure out which moon is which. To help you along, many astronomical periodicals publish charts that identify the relative positions of the satellites from hour to hour and from one night to the next. Several computer software programs, such as SkyMap Pro, also show which satellite is which at any given time. Unfortunately, the rest of Jupiter's family of satellites lie only in the realm of the largest telescopes, so if you see a fifth "moon" near Jupiter, chances are good it's just a background star that happens to be in the same field of view at the moment.

Watching eclipses and, especially, transits of the Jovian moons is also great fun. During a transit, as a moon passes in front of Jupiter, look for its shadow advancing across the cloud tops. You will need a sharp eye and a high-quality, short-focal-length eyepiece to spot it.

## Saturn

FINDING FACTOR:  *
"WOW!" FACTOR:   Binoculars: *      Small telescopes (3″ to 5″): ****
                 Medium telescopes (6″ to 8″): ****

As tantalizing as Jupiter is, it faces some stiff competition from the next planet out. To many, Saturn is the most beautiful sight the heavens have to offer. It's always a crowd pleaser at star parties and public viewing sessions and is probably responsible for luring more people into astronomy than any other celestial object.

In many ways, the planet Saturn is like a miniature Jupiter. Like Jupiter, Saturn is shrouded with a deep atmosphere of hydrogen, helium, and traces of water, methane, and ammonia. Deeper into Saturn's clouds, atmospheric pressure rises rapidly, reaching heights unimagined here on Earth. Below the clouds that we see on Saturn lies a layer of molecular hydrogen surrounding a thick level of metallic hydrogen, which in turn shrouds an Earth-size core of solidified hydrogen rock and ice.

Despite its similar chemistry, Saturn's atmosphere does not show the vivid structure and pastel colors that we see in Jupiter's. Due to the planet's smaller mass (5.68 × 10²⁶ kg versus Jupiter's mass of 1.900 × 10²⁷ kg), Saturn's atmosphere is not heated as greatly as Jupiter's, which is probably the reason for the difference in appearance. Another contributing factor is that, because of Saturn's greater distance from the Sun, solar radiation is not strong enough to create the complex chemical reactions partially responsible for Jupiter's colorful show.

Without a doubt, the single biggest attraction to Saturn is its wonderful system of rings. Galileo saw that there was something peculiar about Saturn's telescopic appearance when he recorded it in 1610, but his telescopes were not sharp enough to make out exactly what was odd about it. His interpretation was further confounded by the fact that Saturn's rings at that time were tilted nearly edge-on as seen from Earth, which would render them nearly invisible even through today's telescopes. Instead, Galileo thought the rings were "handles" or large moons on either side of the planet. His notes recall: "I have observed [Saturn] to be tripled-bodied. This is to say that to my very great amazement Saturn was seen to me to be not a single star, but three together, which almost touch each other." His early drawings of Saturn showed a silhouette almost resembling Mickey Mouse, but with droopy ears!

The mystery was finally solved in 1659, when the Dutch astronomer Christiaan Huygens noted that Saturn is surrounded by "a thin, flat ring, nowhere touching, and inclined to the ecliptic." Today, we know that Saturn possesses an intricate system of rings, all made up of billions of chunks of rock, water ice, and other frozen material, ranging in size from specks to large icebergs several miles across. The rings of Saturn are very broad but very thin. The width of the main rings is about 43,000 miles (70,000 km), but they are only about 12 miles (20 km) thick. While Saturn is 74,855 miles (120,540 km) in diameter at the equator, Saturn's main rings measure 169,880 miles (273,560 km) in diameter.

There are three theories for how the rings of Saturn came to be. One idea suggests that long ago Saturn's strong gravitational forces shattered a small moon in close orbit. A second theory posits just the opposite, suggesting that the material in the rings is so close to the planet that Saturn's gravity prevented it from forming a satellite in the first place. Finally, a third theory has two small moons colliding with each other, with their debris dispersing into a disk around the planet. Which of these, if any, is true is the subject of much speculation among planetary scientists, but we do know that Saturn is not the only planet encircled by a ring system. Jupiter, Uranus, and Neptune are also ringed worlds, although their rings are much too faint to see through amateur telescopes.

### What You'll See

**With the Naked Eye and through Binoculars.** To the eye alone, Saturn looks like a yellowish "star" shining around magnitude 0. While quite bright in its own right, Saturn might be overlooked by beginners who could mistake it for just another bright star. An easy way to tell it apart, even if you can't recognize immediately which is the planet and which is a star, is to follow the rule of thumb that says that stars twinkle but planets do not. Twinkling, technically called scintillation, is the result of starlight being distorted by the turbulence in our upper atmosphere. Common thought is that starlight is more readily distorted because it comes from a point source, unlike the light reflected off a planet, which has a tiny but measurable disk.

Although the rings can be glimpsed through even the smallest telescopes, most binoculars do not have enough magnification to show them. Those with high-power binoculars (i.e., greater than 12×) may be able to see that Saturn appears somewhat oblong or football shaped, but beyond that, there is little evidence of the rings' existence.

Seven-power binoculars, however, can pick out Saturn's largest moon, **Titan.** Titan is unique in the solar system as being the only satellite with a substantial atmosphere. More details are offered below. Look for Titan as a dim, 8th-magnitude speck.

**Through a Telescope.** Many first-time Saturnians don't realize just how easy it is to see the planet's rings with a telescope (Figure 3.7). All it takes is a 2-inch (5-cm) objective magnifying the image 30 times (30×) or so. Unfortunately, many people also believe that higher magnifications will produce a better view. While it is true that the best views of the rings will probably be had between 100× and 200×, the optimum magnification depends greatly on telescope aperture and quality as well as atmospheric seeing conditions. Too much magnification will only spoil the view.

Careful scrutiny through a 3-inch (7.5-cm) or larger telescope will show that Saturn's rings are divided into zones. The most obvious division in the rings is a dark line called **Cassini's Division,** named after the Italian astronomer Giovanni Cassini. The discovery was made in 1675, a decade after his first observation of Jupiter's Great Red Spot. Cassini's Division, separating the fainter, outer **A ring** from the brighter, inner **B ring,** was originally thought to be an empty gap. Photographs taken by the *Voyager* spacecraft in the 1980s, however, show that although it is largely empty, it does contain bits of opaque material. A third major ring, the **C,** or **Crepe ring,** lies just inside the B ring, although even large telescopes have a difficult time making out its dim presence. Try looking for its silhouette against the bright planetary disk. Finally, another feature, called Encke's Minima, appears as a slightly darker zone toward the outer edge of the A ring. Encke's Minima is very difficult to spot even through the largest backyard instruments.

As Earth and Saturn carry on their independent orbits of the Sun, every now and then our world will line up with the plane of the Saturnian rings. When that happens, we see the rings edge-on. Since they are so thin, perhaps only a few thousand feet thick, they effectively disappear from view. That happened last in February 1996. As this book is released, Saturn's rings now show themselves in all their glory, their angular tilt having reached its maximum value of 27° in 2002. That angle will now slowly close every year until

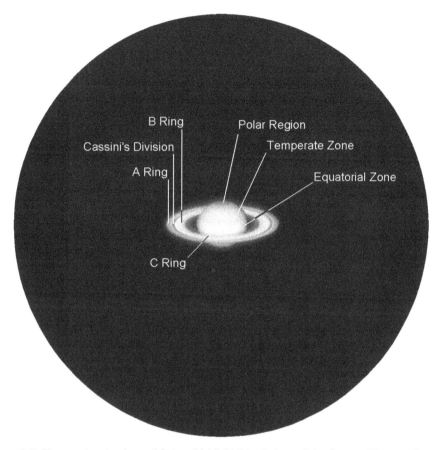

**Figure 3.7** *The spectacular rings of Saturn highlight this photograph by Gregory Terrance through a 16-inch f/5 reflector.*

2009, when they will again appear edge-on in August of that year. Afterward, they will begin to widen once again.

When Saturn is at opposition it appears fully illuminated as seen from our vantage point on Earth, and so appears rather flat, like the Full Moon. Viewing the planet for several months on either side of opposition, however, will easily reveal the shadow of the planet cast onto its rings, creating a spectacular three-dimensional effect. As Saturn reaches opposition, the shadow will shrink and disappear, only to slowly reappear from the opposite edge of Saturn in the days and weeks that follow.

The only banding visible in Saturn's atmosphere through most backyard telescopes is a subtle difference in shading between the **Equatorial Zone, Temperate Zone,** and darker **Polar Regions.** The winds and weather seem much more tranquil than on Jupiter, although approximately once every thirty years, about the same length of time it takes Saturn to circle the Sun, whitish storms develop along Saturn's equatorial regions. The most recent storm, seen in 1990, spread almost completely around the planet. Studies show that Saturn's seasonal storms are apparently caused by large plumes of warmed air rising into the planet's upper atmosphere that are then gradually dissipated by strong equatorial winds.

Of the many satellites that orbit Saturn, none is more intriguing than **Titan,** the second largest moon in the solar system after Jupiter's Ganymede, and the only one to possess an atmosphere. Titan takes a little less than 16 days to complete one orbit of Saturn and is bright enough to be seen with a 2-inch (5-cm) telescope. As many as five or six other satellites of Saturn can be seen through a high-quality 6-inch (15-cm) telescope, although often some are too close to the planet to be spotted. Once again, many astronomical periodicals and software offer charts to aid in the identification of Saturn's moons.

## Uranus

FINDING FACTOR:  ***
"WOW!" FACTOR:    Binoculars: *     Small telescopes (3″ to 5″): *
Medium telescopes (6″ to 8″): **

Astronomical history was made on March 13, 1781, when the renowned astronomer William Herschel accidently stumbled upon a round, greenish object in the constellation Gemini. At first he mistakenly thought that he had discovered a distant comet, for this strange object moved very slowly against the fixed background stars. After analyzing its projected solar orbit however, it was found that Herschel had discovered a new planet out beyond Saturn. That planet later came to be called Uranus, the father of Saturn in Roman mythology.

Uranus, the solar system's third largest planet, measures 31,745 miles (51,120 km) across. Unfortunately, because the planet averages 1.8 billion miles (2.9 billion km) from the Sun, over 19 times farther away than Earth, our view of Uranus has always been severely limited. *Voyager 2*, after gliding past Jupiter in 1979 and Saturn in 1981, reached Uranus in January 1986, becoming the first (and thus far, only) spacecraft to fly past those distant cloud tops. Photographs showed a world completely hidden by uniform clouds of hydrogen and methane; the latter gives the planet its greenish hue. Astronomers believe that Uranus's lack of an internal source of heat results in a stable atmosphere, unlike what we see on Jupiter and Saturn. Additionally, a very hazy stratosphere blocks our view of Uranus's lower troposphere, hiding any structure that might be present.

*Voyager 2* also photographed the planet's tenuous ring system and its five (then-known) icy moons. Perhaps the most impressive find of the *Voyager 2* encounter with Uranus was the discovery of ten additional satellites circling that distant world. Subsequent discoveries have increased the number of natural satellites known to orbit Uranus to twenty-one.

### What You'll See

**With the Naked Eye and through Binoculars.** Although Uranus was the first planet to be discovered in modern times, it was apparently seen by ancient star watchers who plotted it as a dim, stationary point on pretelescopic star maps. Its slow progress across the sky coupled with its dimness must have masked the planet's true nature. Uranus remains visible to the naked eye, although it never shines brighter than 6th magnitude, making it a tough catch even under the best circumstances. Uranus is fairly easy to spot through binoculars if you know exactly where to look. Although it appears only as a starry point, Uranus should still stand out from the crowd of background stars thanks to its unusual, greenish color.

**Through a Telescope.** Telescopes do a little better with Uranus, but even the greatest professional instruments show this distant world as a hazy, featureless disk. Annual finder charts for Uranus are published in the January issues of both *Astronomy* and *Sky & Telescope* magazines. Maps are also found in the Royal Astronomical Society of Canada's *Observer's Handbook* and Guy Ottewell's *Astronomical Calendar*. These star maps will prove indispensable when looking for this elusive planet.

## Neptune

FINDING FACTOR:  ****
"WOW!" FACTOR:    Binoculars: *      Small telescopes (3″ to 5″): *
                 Medium telescopes (6″ to 8″): **

By the beginning of the nineteenth century, astronomers had become mystified by Uranus's behavior. Uranus was the only planet in the solar system that did not follow the path through the sky that the laws of planetary motion predicted it should. Why? One thought was that these irregularities were being caused by the gravitational pull from an unaccounted source, perhaps from another, unknown planet.

Using Newtonian mechanics, two astronomers, John Couch Adams of England and Frenchman Urbain Leverrier, independently calculated how large this undiscovered world must be, how far it was from the Sun, and where it must be in the sky. Leverrier sent positional information to the Berlin Observatory on September 23, 1846, where it was received by Johann Gottfried Galle. Based on this information, Galle and his assistant, Heinrich Louis d'Arrest, discovered Neptune later that same evening using the 9-inch (23-cm) Fraunhofer refractor at Berlin Observatory. Its location in the sky, in the constellation Capricornus, coincided almost exactly with Leverrier's and Adams's predictions, a great triumph for mathematical astronomy.

Our knowledge of Neptune has always been limited by the planet's tremendous distance, an average of 2.8 billion miles (4.5 billion km) from the Sun. All that changed forever in August 1989 when the durable little spacecraft *Voyager 2* flew by Neptune to reveal some amazing details. We now know that the planet is shrouded in a dynamic atmosphere of turquoise clouds. Unlike Uranus, which appears as bland as a cue ball, Neptune's atmosphere is alive with action, perhaps in part due to the planet's internal heat source. The largest atmospheric feature found by the *Voyager 2* cameras was a dark Earth-size vortex that closely resembled Jupiter's Great Red Spot. Appropriately, it was dubbed the Great Dark Spot. But when Neptune was subsequently imaged through the Hubble Space Telescope in June 1994, the Great Dark Spot had vanished. Then in November 1994, a new storm appeared on Neptune's cloud tops, only to fade by the middle of 1995. This would seem to indicate that these storms come and go more quickly on Neptune than Jupiter.

*Voyager 2* also revealed three thin, irregular rings around the planet as well as six new moons to add to Triton and Nereid, two satellites previously discovered through earthly telescopes.

### What You'll See

**Through Binoculars and a Telescope.** Like Uranus, Neptune's challenge to amateur astronomers is simply in finding it, but unlike Uranus, it lies well below naked-eye visibility. In binoculars and telescopes, it appears as a faint, 8th-magnitude starlike point. Of course, there are many stars of similar brightness, but Neptune's unique turquoise color helps it stand out from the crowd. Again, annual finder charts for Neptune are published in the same periodicals as the charts for Uranus.

## Pluto

FINDING FACTOR:   *****
"WOW!" FACTOR:    Binoculars: not resolvable
                 Small telescopes (3″ to 5″): not resolvable
                 Medium telescopes (6″ to 8″): *

We finally arrive at distant Pluto. Dubbed Planet X by astronomers of the late 1800s, Pluto had been the subject of an extensive 25-year-long search before its existence could be confirmed. On February 18, 1930, a young astronomer named Clyde W. Tombaugh discovered Pluto among the stars of Gemini by using a specially designed camera at Lowell Observatory in Flagstaff, Arizona. Though only seen as a dim point of light on his photographic plates, Tombaugh was able to pick out the planet by its slow motion against the starry background.

The orbit of Pluto is inclined 17° to the plane of the solar system, or *ecliptic*, and follows a more elliptical orbit than any other planet. As a result, it is usually far to the north (as it is currently) or far to the south of the Sun's apparent path in our sky. And because Pluto's orbit is so out-of-round, its distance from the Sun can vary greatly, from a minimum of 2.7 billion miles (4.3 billion km) to a maximum of 4.6 billion miles (7.4 billion km). Indeed, one portion of its 248.6-year orbit brings Pluto closer to the Sun than Neptune (as it was from 1979 to 1999).

Pluto is so faint that even the largest telescopes here on Earth record its image as little more than a tiny speck. Since Pluto remains the only planet that has yet to be visited by a spacecraft, we must rely on the Hubble Space Telescope to shed some light on this distant world. We know that Pluto measures only 1,400 miles (2,300 km) in diameter and appears to be a mixture of rock and ice. Its surface is highly reflective, likely covered with a glaze of frozen methane, nitrogen, and carbon monoxide. When closest to the Sun, the slight increase in temperature is enough to partially evaporate some of those frozen gases to form a tenuous atmosphere, only to refreeze as the temperature drops.

Pluto is also circled by a lone moon, Charon, which was discovered in 1978. Charon measures 750 miles (1,200 km) across, creating more of a double planet than a traditional planet-moon system.

### What You'll See

**Through a Telescope.** Even when Pluto is closest to the Sun, as it was in 1989, it is still an extremely faint object. Shining no brighter than 13th magnitude, Pluto requires at least a 6-inch (15-cm) instrument to be seen, and even then, only under exceptionally clear, dark conditions. Most planet watchers agree that an 8-inch (20-cm) is the smallest practical aperture for Pluto hunting. Unfortunately, the Pluto-Sun distance is now increasing, causing its brightness to drop off slightly as the next several years pass.

Detailed finder charts are an absolute must when searching for Pluto. Even then, it is a very difficult challenge to discern which of the multitude of faint specks is actually the planet and not just a random field star. The easiest way is to make a detailed sketch of the star field where Pluto is and then come back a few nights later to see which "star" moved. Once again, check *Sky & Telescope, Astronomy,* the RASC *Observer's Handbook,* or the *Astronomical Calendar* for charts. Use the same technique you did for finding asteroids (see p. 65) by playing connect the dots and looking for distinctive geometric patterns. Good luck!

## Comets

FINDING FACTOR:  * to ****
"WOW!" FACTOR:   Binoculars: Up to ****   Small telescopes (3" to 5"): Up to ****
                Medium telescopes (6" to 8"): Up to ****

Of all the strange and beautiful celestial showpieces that appear in our sky, none have so captured the imagination of mankind as comets. With their long, ghostly tails fanning out across the sky, bright comets immediately draw our attention and awe whenever they flourish.

Once thought to be omens of evil, comets are now known to be members of our solar system. The heart of a comet is the *nucleus,* a conglomeration of water ice and other ices mixed with silicate grains and dust. First proposed in the 1950s by the Harvard astronomer Fred Whipple, this is now known as the "dirty snowball" theory. Dirty iceberg might be more appropriate, since a comet's nucleus averages between 2 and 10 miles (3 and 16 km) in diameter.

Comets orbit the Sun, like the planets. But while the planets follow nearly circular tracks, a comet pursues a highly elliptical path that typically goes from within the orbit of Mercury to well beyond Pluto. The first person to realize this was Edmond Halley, a contemporary of Isaac Newton. In 1705, he published research that traced the orbits of twenty-four comets. He noted that three comets seen previously in 1531, 1607, and 1682 seemed to follow orbits that so closely matched one another that they were likely three appearances of the same comet. Further, Halley predicted that the comet would return in 1758, and while he didn't live to see it, sure enough his comet returned right on schedule. In honor of this discovery, the comet was named Comet Halley. Comet Halley last graced our skies in 1986 and will return in 2061.

As a comet nears the Sun, it begins to warm. Passing the orbit of Mars, water ice in the nucleus begins to *sublimate* (change from a solid directly to a gas), eventually surrounding the nucleus in an expanding cloud called the *coma* or *head*. In the process, some of the silicate grains and dust are also released. The coma can extend for tens or hundreds of thousands of miles around the nucleus.

Larger comets develop tails as they near the Sun. The tail is an extension of the coma, pushed outward by charged particles from the Sun, the *solar wind*. As a result, the tail always points away from the Sun, regardless of the comet's direction of movement. The largest comets may form two tails: a gas tail (sometimes called an ion tail) and a dust tail. In photographs, the dust tail usually appears softer, as the dust particles from the comet fan out along the comet's orbit, while the gas tail is usually straight and narrow, though irregular in structure and texture.

### Where and When to Look

The 2061 return of Comet Halley may seem a long time away, but at any given time, there may be half a dozen or more comets in the sky. Most are only visible in large telescopes and pass unnoticed by most amateurs. At least once or twice each year, a new comet will break the magnitude 9-to-10 barrier and become bright enough to be seen in binoculars. Then there are those comets that seem to make the wait worthwhile. Comet Hyakutake, with one of the longest tails in recent memory, was visible in March 1996, while Comet Hale-Bopp appeared just a year later. Both were striking objects to the naked eye, wonderful to follow night after night as they crept across the sky.

One of the best sources of information on comets that are currently visible comes from the Jet Propulsion Laboratory in Pasadena, California. There Charles Morris maintains the Comet Observation Home Page (encke.jpl.nasa .gov). Updated several times a month, this excellent source lists comets that are currently visible, plots their locations on star maps, and posts photographs and observations submitted by contributors from around the world. Another excellent fount of information is the Central Bureau of Astronomical Telegrams at the Smithsonian Astrophysical Observatory (cfa-www.harvard.edu/ iau/cbat.html), on the grounds of Harvard University in Cambridge, Massachusetts. The Central Bureau is the international clearinghouse for all comet discoveries. If you ever think that you have discovered a comet, they are the folks you want to tell about it first.

### What You'll See

**With the Naked Eye and through Binoculars.** When a comet is predicted to break the binocular barrier (or even more rarely, the naked-eye limit), it receives wide publicity in the astronomical world. To spot it yourself, you will need a decent finder chart and a dark sky, since comets often look like diffuse balls of celestial cotton. Light pollution may brighten the surrounding sky so much that the comet becomes indistinguishable from the background.

Once you've spotted the comet, study its overall appearance. In all likelihood, you will see a bright central core within the comet's coma, where the nucleus is centered. Keep in mind that you are not seeing the nucleus itself, since it is so small, but rather the highest concentration of gases and dust being emitted by the nucleus.

Extraordinary comets may also show a tail extending away from the comet's head. Remember, the tail points away from where the Sun is relative to the comet, so look in that direction. If the comet is visible to the naked eye, it may well have two tails—gas and dust—as was the case with Comet Hale-Bopp in 1998 (Figure 3.8) and, earlier, Comet West in 1976.

**Through a Telescope.** Most comets remain only in the realm of telescopes, and even then they may not brighten to greater than 11th or 12th magnitude. These are of little interest except to devout comet observers. As a comet exceeds 8th or 9th magnitude, however, it becomes bright enough to be visible through smaller amateur telescopes. Again you will need a good finder chart to spot it unless it is very bright. Dark skies are also a must for the best views. Begin with a low-power eyepiece (say, in the 25- to 40-mm range) to locate the comet and to study its overall appearance. Does the comet have a tail? If so, low power is your best bet at seeing it. Are there any knots or other irregularities visible in the gas tail?

Switch to a higher magnification to study the core of the comet's head. On rare instances, you might actually be able to see structure within the coma. Many advanced observers who study comets with great passion use a "Swan-band filter" to enhance contrast and details within a comet's coma.

**Figure 3.8** *Comet Hale-Bopp, the last great comet of the twentieth century, was a bright apparition in the winter skies of 1997. The author took this photograph through a 200-mm lens.*

(A comet's gaseous emissions in its coma and tail fluoresce as sunlight excites molecules of diatomic carbon, which is usually abbreviated "C2" and referred to as the "Swan band." A Swan-band filter enhances the visibility of these gases.)

### Going Further

If keeping close tabs on our fellow members of the solar system appeals to you, consider joining the Association of Lunar and Planetary Observers (ALPO). With members from around the world, ALPO gathers and archives observations from professional and amateur astronomers to help further our knowledge of our solar family. For more information about ALPO programs, contact the membership secretary at P.O. Box 13456, Springfield, IL 62791-3456 or visit the ALPO home page at www.lpl.arizona.edu/~rhill/alpo.

# 4

# The Sun

Like all stars that we see in our sky, the Sun (Figure 4.1) is a huge sphere of hot gases. It is made primarily of hydrogen, heated by a complex thermonuclear fusion reaction that converts hydrogen into helium, and in the process radiates heat, light, and several other types of emissions. Were it not for this solar radiation, life on Earth would not, could not, exist.

Like the planets, the Sun is a member of the solar system. In fact, it holds more than 99 percent of the entire mass in the solar system. Located an average of 93 million miles from Earth, the Sun, a yellow dwarf star, maintains an average surface temperature of 11,000°F (6,500°C), which places it just slightly below the middle of the range in temperatures of the stars that we can see in the universe. Blue giant and supergiant stars can range in surface temperature up to 50,000°F (27,800°C), while red dwarf stars can be as cool as 5,800°F (3,200°C).

The process that produces the heat and energy that radiates from the Sun starts deep within the core, where the element hydrogen is converted into helium in a powerful nuclear fusion process called the "proton-proton cycle." As hydrogen atoms fuse together, energy is generated, obeying the process first formulated by Albert Einstein: $E = mc^2$. Perhaps the most famous scientific equation of all time, this simply says that the amount of energy generated is equal to the mass of the fuel times the speed of light squared. Since the speed of light squared is such a large number, even a tiny amount of mass can produce an amazing amount of energy *under the right conditions.* We can't create those conditions in the laboratory, but it comes naturally to the Sun and the other stars.

The Sun's energy is released in a variety of forms, including intense levels of infrared, visible, and ultraviolet radiation. We must exercise care when viewing the Sun through a telescope or binoculars (or even with the unaided eye, for that matter), since it can harm our eyes. The problem results from a combination of the high-intensity visual light plus the thermal effects from

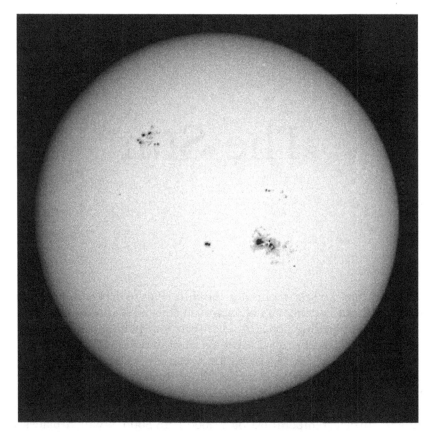

**Figure 4.1** *The Sun, showing a fine collection of sunspots. Jim Fakatselis took this photograph through a 4.1-inch f/6 refractor with a 2× teleconverter for an effective focal ratio of f/12. A Baader Planetarium solar filter was used to reduce the Sun's blinding light to safe levels.*

infrared radiation. If you view the Sun through a telescope or binoculars without proper precautions, then the eye's sensitive retina can be overheated and burned very badly, potentially leading to a permanent blind spot.

I want to emphasize that this is *not* to say that we should shy away from observing our star, just that all observers must approach viewing the Sun with care. Happily, there are several easy and completely safe ways to sun gaze.

"WOW!" FACTOR:  Binoculars:  **      Small telescopes (3″ to 5″):  ****
                Medium telescopes (6″ to 8″):  ****

The simplest way to look at the Sun is not to look at it; instead, use your telescope as a solar projector, as shown in Figure 4.2. By passing the Sun through a telescope or binoculars, its image will be projected out through the eyepiece, just like a slide or movie. Move the screen in and out from the eyepiece to change the size of the Sun's image, and then turn the focusing knob until everything is sharp and clear.

**Figure 4.2** *The author's daughter, Helen, demonstrating how to view the Sun safely by projecting its image onto a screen. While perfectly safe for viewing, solar projection can overheat a telescope, possibly leading to internal damage.*

A smooth, white surface is best for the projection screen, since any irregularities will degrade image quality and lessen the amount of visible detail. Also be sure that the projection screen is not in direct sunlight but rather in shadow, so that the image is not washed out. This shouldn't be a problem for Newtonian reflectors, since their eyepieces stick out to one side, but with a refractor or Cassegrain-style instrument (such as a Schmidt-Cassegrain or Maksutov), it's always best to use a 90° star diagonal to direct the sunlight off to one side of the telescope. When projecting the Sun's image through binoculars, cut out a cardboard light block that fits securely over the outside of the barrels. The light block will cast a shadow onto the screen to keep sunlight from washing out the image. Also, keep one binocular capped to prevent projecting two overlapping images.

One caveat about solar projection. Prolonged viewing can overheat the telescope's and eyepiece's optics, causing damage. Aim at the Sun for no more than ten minutes, then move the instrument off to let it cool. It is best to use inexpensive eyepieces just in case damage results from overheating.

Most seasoned solar observers prefer solar filters to projection, since direct viewing will show the greatest amount of detail. To do this safely, an observer must use a specially designed solar filter that fits over the front of the telescope (or binoculars). In this way, the dangerously intense solar rays are reduced to a safe level prior to entering the telescope and your eyes. But heed this warning: *Never place a solar filter between your eyes and the eyepiece.* The sunlight's magnified heat flooding through the telescope will crack or melt the solar filter in a matters of seconds, exposing your unprotected eye to dangerous levels of solar radiation.

Safe solar filters are made from either glass or a polymer, such as Mylar, and fit securely in *front* of a telescope or binoculars. The most popular solar filters are sold by Baader Planetarium, Thousand Oaks Optical Company, Orion Telescopes, JMB, Kendrick Astro Instruments, and Roger W. Tuthill, Inc. Figure 4.3 shows one example. NEVER view the Sun through photographic neutral-density filters, smoked glass, overexposed photographic film, or other makeshift materials. They can pass invisible ultraviolet and infrared light, causing severe eye damage. Solar filters are discussed in greater detail in my book *Star Ware*.

Finally, never look through the telescope's finder to aim the instrument. In fact, it's a good habit to leave the dust caps on your finderscope just as a reminder. To aim, watch the telescope's shadow as you swing the instrument back and forth. When the shadow is shortest, the telescope is pointed at the Sun.

**Figure 4.3** *Most solar observers prefer to view the Sun through a solar filter mounted in front of their telescopes, as shown here.*

### What You'll See

**Through Binoculars and Telescopes.** For most amateur astronomers, the biggest attraction to observing the Sun is watching the ever-changing array of sunspots across the photosphere. Sunspots were independently discovered by Galileo in Italy, Thomas Harriot in Britain, Johannes Fabricius in Holland, and Christoph Scheiner in Germany, all in 1611. Who among these men actually spotted them first is lost to history, but we do know that Galileo was first to realize that sunspots are physically part of the Sun. Monitoring the progression of sunspots across the solar disk, Galileo noticed that the shape of the sunspots became foreshortened or flattened as they neared the Sun's limb. He astutely concluded that this would only happen if the spots were actually on the surface of the Sun, and not bodies in orbit around the Sun, as others argued. He also found that sunspots take about two weeks to cross the Sun's disk.

Sunspots may look black as coal, but in reality, they are very bright. In fact, if a sunspot could somehow be magically plucked out of the Sun and placed in the night sky, it would shine brilliantly. The largest would even surpass the Full Moon! They only look dark because of their contrast against their surroundings. Sunspots mark cooler regions of the photosphere. In fact, the temperature of a typical sunspot is as much as 2,200°F (1,200°C) less than its surroundings. Though a big difference, that is still hotter than the surfaces of many stars, including Betelgeuse and Antares.

Sunspots are transitory features that develop in unusually strong regions of the Sun's powerful magnetic field. When these regions occur, they block some of the energy pouring outward from the Sun's core, lowering the temperature of the adjacent photosphere.

Each spot has a very dark central portion called the *umbra* (from the Latin for *shadow*) surrounded by a lighter ring called the *penumbra.* Sunspots can range in size from hundreds to thousands of miles in diameter. Some are large enough to be visible through binoculars, and even the unaided eye with proper filtering, but most require a telescope. Sunspots often form in groups, with the largest spanning several Earth diameters across. Typically, groups consist of one or two larger spots surrounded by several smaller members.

Just how many sunspots will be visible at any one time is difficult to predict. Much depends on the optical quality of your telescope and eyepiece, as well as the steadiness of our atmosphere (i.e., seeing). It is best to set up your equipment for viewing the Sun in a grassy yard or field rather than a parking lot, since pavement reradiates image-ruining heat waves. The number of sunspots also depends greatly on where we are in the eleven-year *sunspot cycle.* Toward sunspot maximum, dozens, even hundreds of individual sunspots may be visible, with many clustered into large groups. When nearing sunspot minimum, the solar disk can appear completely barren. The last sunspot cycle peaked in 2001, with the number of spots now on the decline to about 2006, when things should begin to reverse themselves.

Any amateur sun gazer can use the projection method to duplicate the historic observations originally performed by Galileo and determine for him- or herself that the Sun does indeed rotate. Trace a circle on a piece of paper, then draw two lines to divide the circle into four equal segments. Some observers may prefer to use graph paper for even finer division. Hold the piece of paper with a clipboard or other solid backing (an artist's easel is ideal) and move it back and forth from the eyepiece until the Sun's projected image fills the circle. Use a soft-leaded pencil to sketch the locations and appearances of all sunspots that you see, making sure to portray the umbra and penumbra as accurately as you can. Be sure to mark the orientation of the sketch with the four compass directions and note the date and time the drawing was made. Come back and redraw the Sun's disk every one to three days over a span of a month and compare the results. These sketches will reveal how sunspots change in size and shape, some quite remarkably, as time goes by. By extending your observational records over several months, you will also find that spots at different latitudes on the Sun rotate at different speeds. Today, we know that the Sun's equator rotates once every 25 days, while the north and south polar regions take approximately 36 days.

Another solar feature to look for is *faculae*, brighter areas of the photosphere above sunspots. Faculae lie hundreds of miles above the visible photosphere, causing them to be less affected by the dimming effect of the solar atmosphere. Faculae are most easily seen along the edge, or limb, of the Sun, where their contrast is greatest thanks to an effect called *limb darkening*. Limb darkening is caused by our looking diagonally through a greater depth of the Sun's atmosphere toward the disk's limb, or edge, than when we are viewing the central region.

Close-up study of the photosphere will show that it is not smooth but rather composed of myriad tiny grains called *granulation*. Each granule, measuring between 400 and 600 miles (700 and 1,000 kms) across, is a convective cell of heated gas. As the gas rises and reaches the top of the photosphere, it cools, and then drops back into the photosphere to be reheated and recirculated, like bubbles in a pot of boiling water. Since each granule only measures a few arc-seconds across from our distant vantage point, spotting granulation requires at least a 3-inch telescope, 150- to 200-power, and very steady seeing conditions. Less-than-perfect conditions or optics will blur the granules, making them impossible to resolve. Experience has also shown that granulation is difficult, if not impossible, to spot with solar projection.

## Solar Eclipses

Eclipses of the Sun, caused by the New Moon coming between Earth and the Sun, always attract wide attention in the news media as well as from the general public. Even a partial solar eclipse offers a wonderful opportunity to experience the magic of astronomy. The show begins when the Moon takes that first bite out of the Sun's western limb at first contact. Think back to those ancient times when eclipses were seen as being conjured up by angry gods. Then consider how far our knowledge of celestial mechanics has come to be able to predict the coming of an eclipse with such amazing accuracy, centuries in advance.

Always exercise the same precautions when watching a partial solar eclipse as you would when viewing the uneclipsed Sun. Viewing through a telescope equipped with a solar filter, you can watch as the Moon slowly slips across the face of the Sun. With 100× or more, study the silhouette of the rugged lunar terrain's deep valleys and tall mountains, all painted black against the brilliance of the Sun's photosphere. Also keep an eye on any sunspots visible on eclipse day. Watching the slow eastward motion of the irregular lunar limb engulf each spot in its path is fascinating, especially at high magnification. It can be especially fun to follow the Moon as it crosses a large sunspot. First, the spot's lighter surrounding region, the penumbra, is struck by the Moon's pitch-black limb, followed quickly by the sunspot's darker center, the umbra, which first merges into the Moon and then disappears.

If a sunspot or sunspot group happens to lie near the center of the Sun's disk during a solar eclipse, observers can approximate the spot's diameter by using the length of time it takes the Moon to cover it. On average, the Moon will cover the Sun's disk in just under one hour. Since the Sun measures 864,900 miles (1,392,000 km) in diameter, the Moon must pass over 250 miles (400 km) of the Sun's surface every second. Therefore if it takes, say, ten seconds for the Moon to cover a centralized sunspot, the spot must measure close to 2,500 miles (4,000 km) in diameter. If a spot being measured is far from the center of the solar disk, however, the estimate will not be accurate because of the disk's curvature.

Partial eclipses are fun to watch, but it is a total solar eclipse, when the Moon fully blocks the Sun, that calls to observers around the world. Although the Moon is much smaller than the Sun, a mere 2,173 miles (3,476 km) across, it works out that both bodies have a distance-to-diameter ratio of approximately 400 to 1. Put another way, the Sun is about 400 times larger in diameter than the Moon and about 400 times farther away. As a result, the Moon and Sun each appear the same size in our sky, about half a degree.

This near-perfect "fit," with the Moon just covering the Sun's brilliant surface while still exposing our star's chromosphere and its atmosphere, the corona, is critical to the majesty of a total solar eclipse. If the Moon appeared noticeably larger than the Sun in our sky, these characteristics would be blocked from view; if it were smaller, then the bright surface of the Sun would never be fully covered, leaving them lost in the glare. (It turns out that of all the planets and satellites in our solar system, Earth is the only one that enjoys this situation. On all of the other planets, their satellites are either too small or too large to cover the Sun so perfectly.)

The excitement in watching a total solar eclipse really begins to mount when about 85 percent of the Sun is covered and the sky begins to darken. The temperature will begin to fall, and winds may pick up. Animals are fooled into thinking that night is approaching, and many birds go to roost. The brighter planets and stars become visible as the ever-deepening twilight effect encroaches.

In the last few seconds before totality, the Moon's jagged edge breaks up the last remaining rays of sunlight into brilliant luminous fragments, called

*Baily's Beads*. At second contact, the Moon completely covers the Sun. *Only* after the Sun has been *entirely* blocked by the Moon, during the total phase of a solar eclipse, is it safe to view the Sun briefly without protection, since the dangerous rays of the photosphere are blocked. As the last few beads of sunlight disappear, the glow of the Sun's outermost layer, the beautiful corona, blossoms into view, its pearly white glow spreading out for several degrees around the Sun. Figure 4.4 shows totality during the July 1991 total eclipse from Mexico. The corona will remain visible only during the period of totality. The chromosphere, a striking reddish glow encircling the Moon's blackened limb, will also pop into view at second contact. Within the chromosphere are the spectacular flamelike prominences, shooting upward into the corona, only to loop back if their velocity isn't great enough to escape the Sun's tremendous gravity.

The Moon's orbit around Earth is not circular but rather oval or elliptical. At its closest point (called perigee), the Moon is 221,000 miles (356,000 km) away, while at its farthest (apogee), the Moon is 253,000 miles (407,000 km) away. Likewise, Earth's elliptical orbit of the Sun brings it as close as 91,938,000 miles (147,100,000 km) at perihelion (the point where Earth is closest to the Sun), and as far as 95,063,000 miles (152,102,000 km) at aphelion (the point where Earth is farthest from the Sun). As a result, the apparent sizes of the Moon and the Sun vary slightly in our sky. When the Moon appears smaller than the Sun as it passes centrally across the solar disk, a bright ring, or annu-

**Figure 4.4** *A total solar eclipse of the Sun has been described as nature's most dramatic event. This photograph, taken by the author through a 4-inch f/10 refractor, shows the beginning of totality during the July 1991 eclipse over Baja California, Mexico.*

lus, of sunlight remains visible at greatest eclipse. These are called *annular eclipses*. An annular eclipse is also great fun to watch unfold, but keep in mind that a proper solar filter must remain affixed to your telescope and binoculars throughout the entire event.

Table 4.1 lists upcoming solar eclipses between 2003 and 2015.

Total solar eclipses are often described as the universe's most spectacular events. I truly hope that you will be able to see at least one in your life. The next total solar eclipse to cross the United States will take place on August 21, 2017, and April 8, 2024. The 2017 path of totality will pass from Oregon diagonally southeast to South Carolina, while the 2024 eclipse will go from Texas to upstate New York and New England before crossing into eastern Canada. Until then, North America will remain in a total-eclipse drought for the next couple of decades, although the May 2012 annular eclipse will be a wonderful sunset event across the western United States. That annular eclipse's path will cross such landmarks as Bryce Canyon and Grand Canyon National parks. For more information on these and all upcoming solar and lunar eclipses through the year 2017, please refer to my book *Eclipse!* (John Wiley & Sons, 1997).

Table 4.1 **Upcoming Solar Eclipses: 2003–2015**

| Date | Type of Eclipse | Region of Visibility |
|---|---|---|
| 2003 Nov 23 | Total | Australia, New Zealand, Antarctica, southern South America (total eclipse seen in Antarctica) |
| 2004 Apr 19 | Partial | Southern Africa, Antarctica |
| 2004 Oct 14 | Partial | Hawaii, Alaska, northeastern Asia |
| 2005 Apr 8 | Annular/Total[1] | North America, South America (maximum eclipse seen in Panama, Colombia, Venezuela) |
| 2005 Oct 3 | Annular | Europe, Africa, southern Asia (annular eclipse seen in Portugal, Spain, Libya, Sudan, Ethiopia, Kenya) |
| 2006 Mar 29 | Total | Africa, Europe, western Asia (total eclipse seen in central Africa, Turkey, Russia) |
| 2006 Sep 22 | Annular | South America, western Africa, Antarctica (annular eclipse seen in Guyana, Suriname, French Guiana, South Atlantic) |
| 2007 Mar 19 | Partial | Alaska, Asia |
| 2007 Sep 11 | Partial | South America, Antarctica |
| 2008 Feb 7 | Annular | Antarctica, eastern Australia, New Zealand (annular eclipse seen in Antarctica and adjacent Indian Ocean) |
| 2008 Aug 1 | Total | Northeastern North America, Europe, Asia (total eclipse seen in northernmost Canada, Greenland, Siberia, Mongolia, China) |
| 2009 Jan 26 | Annular | Southern Africa, Antarctica, southeastern Asia, Australia, (annular eclipse seen in southern Indian Ocean, Sumatra, Borneo) |

*(continued)*

*Table 4.1* **Upcoming Solar Eclipses: 2003–2015 (continued)**

| Date | Type of Eclipse | Region of Visibility |
| --- | --- | --- |
| 2009 Jul 22 | Total | Hawaii, eastern Asia, Pacific Ocean (total eclipse seen in India, Nepal, China, central Pacific) |
| 2010 Jan 15 | Annular | Africa, Asia (annular eclipse seen in central Africa, India, Myanmar, China) |
| 2010 Jul 11 | Total | Southern South America (total eclipse seen in South Pacific, Easter Island, Chile, Argentina) |
| 2011 Jan 4 | Partial | Europe, Africa, central Asia |
| 2011 Jun 1 | Partial | Eastern Asia, northern North America, Iceland |
| 2011 Jul 1 | Partial | Southern Indian Ocean |
| 2011 Nov 25 | Partial | Southern Africa, Antarctica, Tasmania, New Zealand |
| 2012 May 20 | Annular | Asia, Pacific, western North America (annular eclipse seen in western United States, China, Japan, Pacific Ocean) |
| 2012 Nov 13 | Total | Australia, New Zealand, South Pacific, southern South America (total eclipse seen in northern Australia, South Pacific) |
| 2013 May 10 | Annular | Australia, New Zealand, central Pacific Ocean (annular eclipse seen in northern Australia, central Pacific) |
| 2013 Nov 3 | Annular/Total | Eastern North and South America, southern Europe, Africa (maximum eclipse seen in Atlantic Ocean, central Africa) |
| 2014 Apr 29 | Annular | Southern Indian Ocean, Australia, Antarctica (annular eclipse seen in Antarctica) |
| 2014 Oct 23 | Partial | North Pacific Ocean, North America |
| 2015 Mar 20 | Total | Iceland, Europe, northern Africa, northern Asia (total eclipse seen in North Atlantic, Faeroe Island, Svalbard) |
| 2015 Sep 13 | Partial | Southern Africa, southern Indian Ocean, Antarctica |

1. *An annular/total eclipse occurs when the size of the Moon's disk is exactly equal to the size of the Sun's disk as seen in our sky. As a result, totality will last just an instant along the path of maximum eclipse.*

# 5

# The Deep Sky

For the purposes of this book, let's imagine that the universe is divided in half, separated by an invisible wall that lies beyond the orbit of Pluto. Inside that imaginary boundary lies what some call the *shallow sky*. That's where we find the Sun, the Moon, all of the planets, asteroids, meteoroids, and comets—in short, the solar system. As we have already seen in earlier chapters, each member of this elite group has a unique place, each with its own special features, which are a source of endless fascination. But all of these are in our celestial backyard. It's time to explore the rest of our cosmic neighborhood—and beyond!

Past the solar system lie the depths of the universe, the *deep sky*, where we find the stars that light the night as well as other treasures, often too faint to be seen with the eye alone. Amateur astronomers often call these *deep-sky objects*. This chapter will examine and explain the different types of deep-sky objects, while the next four chapters will highlight the finest within the four seasonal skies.

Many new stargazers begin their deep-sky experience by aiming at the brighter stars in the sky, only to learn that those stars remain as no more than points of light. While it is true that most of the night's brightest stars are bigger and brighter than our own Sun, they also lie at tremendous distances. In fact, they are so far away that stars will only show as points of light regardless of telescope size or magnification, rendering them of little interest other than perhaps for a passing glance. (It is true that, under high magnification, a star will show a tiny disk, but this is caused by its light interacting with a telescope's optics and is not the star's true disk.)

Instead of spending time looking at individual stars, we will concentrate on more exciting types of deep-sky objects. And there are hundreds, even thousands, of double and variable stars, star clusters, nebulae, and galaxies visible through backyard telescopes. All it takes is a little time and effort to find them.

Here are general descriptions of the major classes of objects. Facts about specific examples of each may be found in later chapters.

## Double Stars

Appearances to the contrary, many of the stars we see are not single points of light but instead are two or more closely set stars in orbit around each other. These are called either *double stars* or *binary stars* (Figure 5.1), or if more than two suns are involved, *multiple stars*. The brightest star in a double-star system is always referred to as the primary or "A" star, and the fainter companion is always the secondary or "B" star. If there are more stars involved, the others will be assigned letters in descending order, such as "C," "D," and so on.

The gap between two members of a double-star system is usually measured in arc-seconds, although some wider pairs may be specified in arc-minutes. Of course, in reality the stars may be millions or billions of miles apart.

If we happen to be seeing the star system edge-on, then the system's total light output will dip every time the companion sun passes in front of or behind the primary star. These are called *eclipsing binaries*. When both stars are visible side by side, their collective brightness is greatest. As the companion star passes behind the primary in eclipse, the system's magnitude drops

**Figure 5.1** *The double star M40 in Ursa Major, which is highlighted in Spring Sky Window 2. George Viscome took this photograph through a 14.5-inch f/6 reflector. South is positioned at the top to reflect the view through most astronomical telescopes.*

off. The light output returns to normal when the eclipse ends, only to lower a second time as the secondary sun passes in front of the primary.

Here is some good news for city dwellers and others impacted by light pollution. Since double stars are point sources of light, they do not suffer as badly from the ills of civilization as many other deep-sky objects. This makes them great targets for urban astronomers as well as for nights when the Moon shines brightly.

## Variable Stars

While we normally think of stars as unwavering in nature, many appear to change in brightness over periods that range from a few hours to several years. Some do so with great regularity, while others act erratically. Collectively, these are known as *variable stars*. Variable stars can be divided into three categories, depending on their behavior. The first, eclipsing binaries, was already discussed under "Double Stars."

The most common type of variable stars are *pulsating variables*, stars that actually expand and contract in size. *Long-period* pulsating variables take several months to complete their cycles, going from maximum brightness to minimum light output and then back to maximum again with great regularity. There are many long-period variables scattered throughout the sky, with the best known being Mira, a star in the autumn constellation Cetus (see chapter 8). Other types of pulsating variables have comparatively short periods. One of the most interesting short-period variables to study through binoculars and backyard telescopes is Delta Cephei, a yellow star found in the constellation Cepheus. Cepheus is a circumpolar constellation for most of the midnorthern latitudes, though it is most easily visible in the autumn. As chapter 8 discusses in greater depth, Delta Cephei is the archetypical model of a group of variable stars called Cepheids. Studies show that a Cepheid's period of variability is proportional to its inherent luminosity.

*Erupting variables,* as the name implies, are stars that spend most of their time shining at a fixed brightness then suddenly rise several magnitudes. They remain at peak brightness for several days and then slowly fade back to their preoutburst magnitudes. Erupting variables include the explosive novae and supernovae.

## Star Clusters

*Open star clusters* (Figure 5.2) are groups of young, hot stars held together by mutual gravitational attraction. Sometimes called galactic star clusters, each may contain anywhere from a few to a few hundred individual stars. All of the stars in a given cluster lie about the same distance away from Earth and are all about the same age. Astronomers believe that all of the stars in an open cluster were formed from a common molecular cloud of interstellar gas and dust called a *Hydrogen-II region* (discussed below).

Open star clusters and Hydrogen-II regions are both found along the plane of the Milky Way galaxy, as well as in other galaxies. In fact, there are

**Figure 5.2** *Open star cluster M52 in Cassiopeia, positioned within Autumn Sky Window 1. South is up in this photograph taken by George Viscome through a 14.5-inch f/6 reflector.*

open star clusters forming right now, allowing astronomers to study the process of stellar birth firsthand. For instance, the clouds of the Orion Nebula (M42) engulf a forming star cluster. The stars themselves are not visible to the eye directly, since they are hidden behind the nebula's opaque clouds. But by using special, infrared equipment attached to their telescopes, astronomers can count over one thousand members in the group.

Though they travel through the galaxy together for the moment, the stars in an open cluster will eventually scatter over eons of time. As they drift along their orbits, some of their members escape the cluster due to changes in velocity after passing near another cluster star or a particularly dense area of nebulosity, or because of shifts in tidal forces within the cluster's gravitational field.

*Globular star clusters* (Figure 5.3) are huge, spherical agglomerations of stars that surround the Milky Way's galactic nucleus as well as those of other galaxies. Each globular contains between 10,000 and 1 million stars.

Although globular clusters possess many more stars than open clusters, it is much harder to see the individual stars through backyard telescopes because of their greater stellar densities and vast distances from Earth. The densities of globular clusters are measured on a 12-point scale of classes. Class 1 globulars are very compressed, while Class 12 globulars are quite loosely structured. Of the twenty-nine globulars described in later chapters, M55 in summertime's Sagittarius is the least concentrated, rating an 11 on the scale. By con-

**Figure 5.3** *Globular star cluster M12 in Ophiuchus, found within Summer Sky Window 3. George Viscome took this photograph through a 14.5-inch f/6 reflector. South is up and to the right.*

trast, M75, also found in Sagittarius, is so densely packed that it is rated as Class 1. Both are described in chapter 7.

Still, looks aside, globular clusters are mostly empty space. Even at the cores of the densest globulars, there are still many light-months between stars. That's far enough away that the stars would still appear as points of light from an imaginary planet orbiting one of the cluster suns. But imagine what it would be like to live on a planet orbiting one of the stars within a globular cluster! That is just what Isaac Asimov did in his book *Nightfall*, a story about a planet orbiting a six-star system within a globular cluster. And we think that we have trouble with light pollution!

## Nebulae

The word *nebula* (plural: *nebulae*) comes from the Latin word for *cloud*. Astronomers used to call everything beyond the solar system that was not a star a nebula. The advent of the telescope cleared up some of the discrepancy, although it was not until people began photographing distant stellar objects that many star clusters and galaxies were correctly identified. Now when astronomers refer to a nebula, they are talking about an interstellar cloud of primarily hydrogen gas and dust particles. There are several types of nebulae in the universe: bright nebulae, which may be broken down into several subcategories;

planetary nebulae; and supernova remnants. Although all three are called nebulae, they have far different natures.

*Bright nebulae*, among the most beautiful objects in the night sky, can be thought of as stellar maternity wards, for it is inside these giant clouds that new stars are born. Some of the most breathtaking astrophotographs ever taken show soft, gossamer tendrils of nebulosity swirling and floating in interstellar space, often encircling stars in their ghostly grip. Perhaps the most famous bright nebula of all is M42, the Orion Nebula, in the winter sky. Other standout examples include M8 and M17, both in summertime's Sagittarius.

If the level of ultraviolet energy from the young stars within the nebula is strong enough, it ionizes the cloud's hydrogen gas by severing the atomic bond between the hydrogen's nucleus (a single proton) and the orbiting electron. This release of energy causes the nebula to fluoresce, much like the gas in a neon sign glows as electrical current passes through. Bright nebulae of this sort are also called *emission nebulae* or Hydrogen-II regions. Long-exposure photographs show Hydrogen-II regions as brilliant red, the color emitted by ionized hydrogen.

Many emission nebulae, notably M16 in the summer constellation Serpens, as well as M42, M8, and M17 (Figure 5.4), contain pockets of condensing gases, sometimes called EGGs (for evaporating gaseous globules), where new stars are being formed. Over millions or perhaps billions of years, the energy from these newly forming stars, created in their cores by hydrogen atoms fusing into helium, will cause the leftover nebulosity to dissipate, leaving in its wake an open star cluster. Just how many stars will be formed from a Hydrogen-II region depends on the mass contained in the original nebula as well as on other mitigating factors, such as external gravitational forces.

Other bright nebulae shine because the light from stars that are embedded in them is reflected off of dust particles in the nebula. Astronomers refer to these clouds as *reflection nebulae*. Reflection nebulae, composed of uncountable particles of interstellar dust, tend to absorb the red end of the spectrum of light emitted from their illuminating stars, causing them to appear a deep blue in color photographs. The most obvious reflection nebulae visible through backyard instruments is M78, a small tuft of cloudiness in Orion the Hunter.

Two other varieties of nebulae are associated with stellar death. As a typical star, such as our Sun, goes through its death throes, its interior will begin to contract under gravity. During this process, the star's outer layers expand and cool, causing the star to become a red giant. Eventually, the outer layers separate from the inner core and begin to expand into space to form a *planetary nebula*, or *planetaries*, for short (Figure 5.5). Just how long this process takes depends on the initial mass of the star. The larger the star's mass, the faster the process. For instance, a star with the same mass as our Sun will take about 1.5 billion years, while a star with three times the mass will take about 200 million years. Stars with less mass than our Sun will take several billion years to evolve, though they may not possess enough mass to create a planetary. Most planetaries last less than 50,000 years (a blink of the eye on the cosmic time scale) before their shells dissipate.

**Figure 5.4** *The bright nebula M17, known as the Horseshow or Swan Nebula, in Sagittarius is found in Summer Sky Window 5. South is up in this photograph by George Viscome through a 14.5-inch f/6 reflector.*

## How a Star Is Born

Scattered across the outer periphery of spiral galaxies are large, swirling clouds of gas (mostly hydrogen) and dust called nebulae. As a nebula continues to churn, pockets of denser material can gradually form over eons of time. Slowly, gravity pulls in more and more material. The greater the nebula's mass, the greater the gravitational collapse. If enough mass is gathered, a spheroidal object is formed that astronomers call a globule. Continuing to collapse, a process that can take between 10,000 and 1 million years, the globule begins to emit infrared radiation. This, combined with external forces from the surrounding nebula, also causes the globule to begin to rotate.

As the collapse continues, the globule's internal temperature, pressure, and rotation rate increase as its atoms are crushed closer together. Centrifugal forces sculpt the globule into a central core, called a *protostar,* surrounded by a flattened disk of dust, called a *protoplanetary disk.* The dense, central core continues to evolve toward stardom while the protoplanetary disk may eventually evolve into a family of orbiting planets, moons, and asteroids, like our solar system.

If the protostar has sufficient mass and reaches a sufficient internal temperature, nuclear fusion begins. Hydrogen atoms are fused together to form helium, releasing energy in the process. A state of gravitational equilibrium is reached, halting the inward contraction. A star is born.

**Figure 5.5** *Planetary M27, the famous Dumbbell Nebula, is located in the constellation Vulpecula, within Summer Sky Window 7. South is up. George Viscome took this photograph through a 14.5-inch f/6 reflector.*

The term *planetary nebula* is a wonderful example of tradition clouding present knowledge. Astronomers of the eighteenth and nineteenth centuries thought these expanding shells of gas looked like planets similar to the then recently discovered planets Uranus and Neptune. Perhaps the best-known planetaries are M57, the famous Ring Nebula, and M27, the Dumbell Nebula, both of which are visible in the summer sky. Many appear bluish or greenish in telescopes, testifying to the fact that they contain ionized oxygen, which is known to glow with a turquoise hue under certain conditions.

As the universe's most massive stars age, their cores continue to increase in density, causing helium atoms to fuse into carbon and, subsequently, into heavier elements still, such as oxygen, neon, sodium, sulfur, and even iron. As fusion continues inside a massive star's core to the point of creating iron, the core collapses, causing the star to become inherently very unstable, leading to the star's end.

When massive stars die, they erupt in violent explosions called *supernovae*. Supernovae come in two very different varieties, known as Type I and Type II. A Type I supernova only occurs when a white dwarf is teamed with a companion star in a close binary system. As mass is transferred from the companion star to the white dwarf, so a layer of hydrogen accumulates on the dwarf's surface. If the white dwarf's mass exceeds 1.4 times the mass of our

### How Stars Die

Most stars enjoy relatively stable lives for millions, even billions, of years, with the most massive stars having the shortest lives. The Sun, a middle-of-the-road kind of star, has a life expectancy of about 10 billion years, of which about half has been used up.

Over time, as a star begins to use up its supply of hydrogen in its core, the star will begin to expand. In the case of lower-mass stars like the Sun, the star will expand and cool, becoming a red giant star, perhaps 100 times its original diameter. Once all of the hydrogen has been fused into helium, the helium-rich core will contract to create a white dwarf, while the outer layers of the star will be exhausted in a large enveloping cloud called a planetary nebula. The star itself continues to shrink, raising its temperature in the process and probably causing one or more outbursts called *novae* (singular: *nova*). The dying star will continue to radiate leftover heat for millions, even billions, of years but will eventually fade away as a dark stellar corpse, a carbon-rich black dwarf.

When the most massive stars run out of hydrogen in their cores, their outer diameters expand to enormous proportions. Called *red supergiants*, stars at this stage measure about 500 million miles across. As the outer surface expands, the core shrinks, becoming hotter and denser. As the temperature rises, nuclear fusion of helium occurs, creating carbon and oxygen. Again, the core shrinks, the temperature rises, and carbon and oxygen are subsequently fused into sodium. This cycle of contraction, heating, and fusion is repeated several times, creating heavier elements in the process. Finally the star's core becomes composed largely of iron.

Nuclear fusion stops at this point. The core becomes very unstable, triggering an all-consuming explosion called a supernova. When the explosion clears, all that is left is an expanding cloud of debris and the extremely dense core of this once mighty star. If the star's core is between about 1.5 and 3 times the mass of the Sun, it will collapse into a small, dense neutron star about ten miles in diameter made up entirely of neutrons. Neutron stars spin extraordinarily fast and have very strong magnetic fields. If the remaining mass is greater than three times the mass of the Sun, the star literally turns itself inside out and becomes a *black hole*, an incredibly dense object with a gravitational field so strong that not even light can escape.

Sun (what is called the *Chandrasekhar limit*), the white dwarf literally crushes itself, leading to an all-consuming thermonuclear explosion.

A Type II supernova occurs when the iron-rich core of one of the most massive stars becomes unstable. As enormous temperatures and pressures develop, the core collapses rapidly, triggering the supernova detonation.

In either case, when the blast from a supernova finally clears, all that remains of the exploding star is its dense central core and an expanding nebula of gaseous debris called a *supernova remnant*. M1, the famous Crab Nebula in Taurus the Bull, is the sky's most easily seen supernova remnant. It is discussed in chapter 9. (Note that the final path a star takes toward its death depends on how much mass it has when it is ready to die, not the mass when it was formed. Some stars born with more than five times the mass of our Sun will still follow the Sun's evolutionary path, provided they lose enough mass during their lifetimes to trim down to at most 1.4 times the Sun's mass by the time they run out of fuel to power their internal thermonuclear reactions.)

## Galaxies

From extensive studies of photographic plates taken in the early part of the twentieth century, the American astronomer Edwin Hubble concluded that galaxies are vast independent stellar systems, not just nebulae, as had been previously assumed. He further demonstrated that all galaxies fit into three major classes according to their physical shape.

Astronomers classify the Milky Way and other similarly shaped galaxies, characterized by a central core and two or more spiral arms, as *spiral galaxies* (Figure 5.6). Spiral galaxies may be broken down further based on how tightly wound the arms appear. Those with very tight arms are categorized Sa. Others that are less tightly wound are termed Sb and Sc, while the loosest are Sd. The Andromeda Galaxy, M31, is listed as an Sb spiral, while nearby M33 in Triangulum is an Sc spiral, indicative of a looser structure. I must be quick to point out, however, that neither will show its spiral structure through most backyard telescopes.

Rather than curling directly away from the core, the arms of some spirals expand from the ends of unusual barlike features that protrude from their cores. These are called *barred spirals,* and are categorized using the same 4-point rating scale as above, but with the addition of a capital B inserted (e.g., SBa, SBb, etc.). M95 in Leo, for instance, is an SBb barred spiral. Some evidence suggests that our Milky Way galaxy may be a barred spiral.

**Figure 5.6** *Barred spiral galaxy M61 in Virgo can be found within Spring Sky Window 5. George Viscome took this photograph through a 14.5-inch f/6 reflector. South is toward the top.*

Other galaxies bear no resemblance to the spirals at all. Instead they look like enormous oval and circular spheres of stars. These are called *elliptical galaxies* and are the most common type of galaxy in the universe. M32 and M110, both satellite galaxies of M31, are elliptical galaxies. Like spirals, ellipticals are also classified according to appearance and structure, but on a 7-point scale. An E0 galaxy is seen as a perfectly round sphere, an E7 galaxy is flattened strongly at either end, while intermittent points on the scale indicate varying degrees of ellipticity. Strangely, unlike spiral galaxies, elliptical galaxies do not contain any hints of nebulosity or regions of star formation.

Finally, galaxies that do not fit into either the spiral or the elliptical families are termed *irregular galaxies*. As the name implies, irregular galaxies share no common shape. Instead most appear to be amorphous glows. M82 in Ursa Major is a good example of an irregular galaxy.

Examples of each of these types of deep-sky objects can be found in each of the next four chapters. Begin your tour of the universe with tonight's sky and continue throughout the year, journeying to far-off vistas without ever leaving your backyard.

# 6

# Spring Sky Windows

The stars that make up our springtime night sky (Figure 6.1) are like a trip back home for many amateur astronomers. How many of us, as we were growing up, learned about the Big Dipper, that familiar pattern of seven stars that rises high in the sky? You'll find it up again in the spring, welcoming us back to a place that we may not have visited recently. Joining the Big Dipper are several other stars and constellations that we will use to probe the farthest depths of intergalactic space. While the plane of our galaxy, the Milky Way, dominates both the summer and winter skies, our spring window on the universe opens toward myriad distant galaxies, each a separate system made up of billions of stars.

Let's begin our tour with an old friend, the Big Dipper, which is formed by four stars (Megrez, Dubhe, Merak, and Phecda) marking its cup and three stars (Alioth, Mizar, and Alkaid) denoting its crooked handle. If you live in Europe, then you probably know this pattern as the Plough.

The Big Dipper is not an official constellation but instead is part of a much larger group named Ursa Major, the Great Bear. The bowl of the Dipper forms the Bear's back and belly. Faint stars extending to the west of the bowl (or to the left as seen on Figure 6.1) make up the Bear's neck and head, while others curving below (to the south) form its legs and paws. The three stars in the Dipper's handle create the Bear's tail, or, as in Native American legend, three hunters chasing after the bear.

Draw a line between the two stars at the end of the bowl (Merak and Dubhe) and extend the line to the north. The dim star that the line points toward is the North Star, Polaris. Many people are under the mistaken impression that Polaris is the brightest star in the night sky. Not even close! In fact, the North Star is only the forty-ninth brightest star in the sky. What makes the North Star special is that, right now, Earth's north polar axis is aimed almost directly at it. As a result, Polaris appears fixed in the night sky while all other stars appear to rotate around it. For centuries, Polaris has helped people navigate by the stars.

# Spring sky

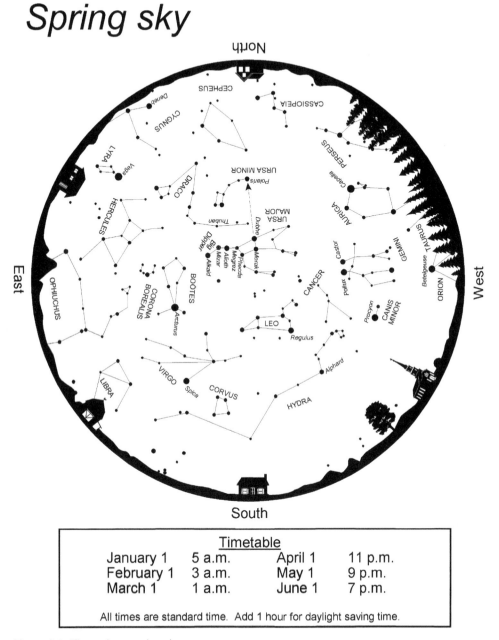

North

CASSIOPEIA

CEPHEUS

Deneb

CYGNUS

PERSEUS

LYRA

Vega

Capella

DRACO

URSA MINOR

Polaris

AURIGA

HERCULES

Thuban

URSA MAJOR

Dubhe

TAURUS

Betelgeuse

Merak

Castor

GEMINI

Big Dipper

Mizar

Alioth

Phecda

Alkaid

Megrez

ORION

CORONA BOREALIS

BOÖTES

CANCER

Pollux

Procyon

OPHIUCHUS

Arcturus

LEO

CANIS MINOR

Regulus

East

West

VIRGO

Spica

CORVUS

Alphard

LIBRA

HYDRA

South

| Timetable | | | |
|-----------|------|--------|----------|
| January 1 | 5 a.m. | April 1 | 11 p.m. |
| February 1 | 3 a.m. | May 1 | 9 p.m. |
| March 1 | 1 a.m. | June 1 | 7 p.m. |

All times are standard time. Add 1 hour for daylight saving time.

**Figure 6.1** *The spring evening sky.*

Polaris is at the end of the handle of the Little Dipper. Just as the Big Dipper is part of larger Ursa Major, the Little Dipper is part of Ursa Minor, the Small Bear. The seven stars that mark Ursa Minor are all quite faint. Suburban amateur astronomers often check to see how many of these stars are visible to gauge the clarity of the night sky. If the entire Little Dipper can be seen, then the sky is quite clear and dark.

Return to the Big Dipper's bowl and look at the other two stars in the bowl, Megrez and Phecda. Follow the line they make toward the southwest to locate the star Regulus, the brightest star in Leo the Lion. Regulus marks the Lion's heart. Above Regulus are five stars that combine to form Leo's head (they look like a backward question mark or a sickle). The body of Leo stretches out to the east, with the animal's hindquarters and tail represented by a triangle of three stars.

Continue the line from the Dipper through Regulus and farther toward the southwest. Look for the bright star Alphard and its constellation, Hydra, the Water Snake. From head to tail, Hydra stretches a full third of the way around the sky, making it the longest constellation of all. Apart from Alphard, Hydra's stars are very faint, making it difficult to trace the full length of this constellation.

Return to Regulus momentarily, then look toward the northwestern horizon for a pair of bright stars. Those are Castor (on the right, or north) and Pollux (on the left, or south), twin stars in the constellation Gemini, the Twins, which is best seen during the winter months. Halfway in between Leo and Gemini are the five faint stars that form the inconspicuous constellation Cancer the Crab.

Let's head back to the Big Dipper once again. Draw a curved line through the three stars that form its handle and continue the arc counterclockwise until you "arc to Arcturus," an orange-colored star that is the brightest star of the season. A zero-magnitude star, Arcturus belongs to the constellation Boötes the Herdsman. Our modern-day minds may have a hard time imagining a human form among the stars of Boötes. Most people picture it as a kite, or even an ice cream cone. If you look carefully, you just might spot the smile on a child's face who is about to lick the ice cream. That smile, formed by a semicircle of stars to the east, is another constellation, Corona Borealis, the Northern Crown.

Retrace the "arc to Arcturus" then continue to "speed on to Spica." Take a careful look at the colors of both Spica and Arcturus, and note their difference. As mentioned in chapter 4, a star's color depends on its temperature. Spica's blue-white tint tells astronomers that it is much hotter than orange Arcturus. Spica belongs to the zodiacal constellation Virgo the Maiden.

One more time, return to the Big Dipper's handle. Follow the "arc to Arcturus," "speed on to Spica," then continue the "curve to Corvus," the Crow. Even though its stars are faint, Corvus is surprisingly easy to spot low in the southern sky, riding on the back of Hydra. To see a crow here, however, do not think of a trapezoidal pattern as shown on the star map. Rather, imagine a cross tilted on one side. The Crow's body extends between the upper right (northwest) and lower left (southeast) stars. Its outstretched wings spread from the upper left (northeast) to the lower right (southwest) stars.

Figure 6.2 holds the key to the spring sky windows. The map, a duplicate of Figure 6.1, shows seven highlighted areas that are discussed in detail later in the chapter. Each of those windows shows an enlarged area of the sky plotting fainter stars and selected deep-sky objects. Compare the two maps, decide which area you wish to explore tonight, and then head outside to find that region. Once the naked-eye stars and constellations have been located, switch to the close-up sky window found later in this chapter and, using your finderscope or binoculars, begin your journey into the depths of the universe.

Each sky window includes a miniature all-sky chart to show its position relative to the rest of the night sky. To help you better appreciate the scale of things, a crosshair frame is also included in each window to show the typical field of a 6 × 30 finderscope (about 7°).

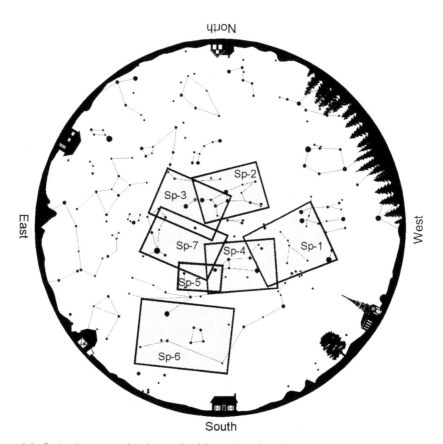

**Figure 6.2** *Spring key map, showing each of the spring sky windows to come.*

## Spring Sky Window 1

Spring Sky Window 1 (Figure 6.3) is framed by the twin stars Castor and Pollux in Gemini to the west and the westernmost portions of Leo the Lion to the east. Within this frame lie the faint stars of Cancer the Crab, although trying to spot the dim form of the Crab usually proves to be an exercise in futility. Although not much to look at with the eye alone, Cancer includes several interesting telescopic objects, including one of the first discovered deep-sky objects.

### M44 (NGC 2632): Open Star Cluster in Cancer

NICKNAME: Beehive Cluster or Praesape
DISTANCE FROM EARTH: 577 light-years
FINDING FACTOR: *
"WOW!" FACTOR:    Binoculars: ****    Small telescopes (3″ to 5″): ***
                 Medium telescopes (6″ to 8″):  **

#### Where to Look

M44 is located in the center of the faint constellation Cancer. While none of the stars in Cancer shines brighter than magnitude 3.5, the bright winter stars Castor and Pollux to its west and bright star Regulus to the east help point the way. Center your finderscope or binoculars about halfway in between Pollux and Regulus. Look for the stars Asellus Borealis (which translates to "northern donkey") and Asellus Australis ("southern donkey"), belonging to the faint body of the Crab, and in between is the dim glow of open star cluster M44. M44 might be visible with the unaided eye from moderately dark suburban locations and almost certainly from the country. The cluster is obvious with even the slightest optical aid.

#### What You'll See

**Through Binoculars.** Nicknamed Praesape (Latin for *manger*) or the Beehive Cluster, M44 comes alive in sparkling style through just about any pair of binoculars, no matter how large or small, expensive or not. Up to thirty stars between 7th and 9th magnitude are seen scattered across the cluster through 7× binoculars, with some of the points set in interesting pairs or patterns. Thanks to its barren surroundings, M44 really dazzles the eye.

**Through a Telescope.** While a couple of dozen stars can be seen in M44 through binoculars, fifty or more stars are resolved through backyard telescopes (Figure 6.4). Most appear white or slightly bluish, although four show a distinctly orangish or reddish tint. Be sure to use your lowest-power, widest-field eyepiece for the best view. Even then, depending on your telescope, you may not be able to fit the entire cluster inside the eyepiece's field of view because of its wide span, 1.5°. That's the same as three Full Moons stacked end to end. The saying "you can't see the forest for the trees" certainly applies to widespread

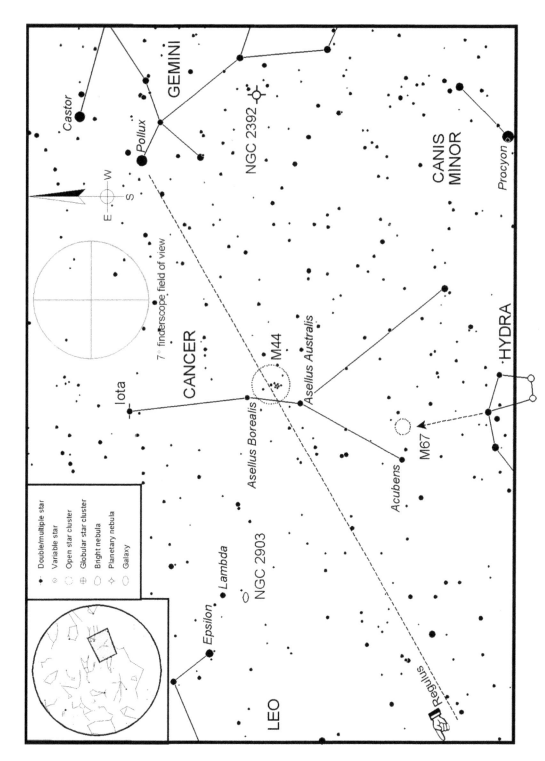

**Figure 6.3** *Spring Sky Window 1, showing the constellation Cancer framed by Gemini to the west and Leo to the east.*

**Figure 6.4** *M44, the Beehive Cluster or Praesape. This photograph taken by George Viscome through a 14.5-inch f/6 reflector is oriented with north at the top to match the view through binoculars.*

open clusters like M44. In fact, you may find that the best view of M44 is actually through binoculars or your finderscope.

One of the most interesting telescopic sights to look for within M44 is a trio of closely set stars listed in catalogs as Burnham 584. These three almost equally bright points of light may be found forming a small triangle just south of the cluster's center.

### M44

M44 was first recorded by Aratos as a "little mist" as far back as 260 B.C. The true nature of the Praesape came to light when Galileo viewed it through his first crude telescope. Today, even the smallest field glass will resolve many of its member stars.

Over two hundred stars belong to M44. Of these, most are white, spectral-type A stars, with about seventy-five shining brighter than 10th magnitude. As mentioned above, at least four other cluster stars glisten with contrasting reddish or orangish tints. Astronomers know that these four stars are the most massive within M44. How? The greater a star's mass, the faster it evolves. These four red stars have already evolved past their main-sequence stage to become red giants. By studying the different types of stars within a star cluster like M44, astronomers can gain better insight into each member sun's individual age, as well as the cluster's overall age. Based on these studies, M44 is thought to be about 400 million years old, still an infant compared to our 4.5-billion-year-old Sun.

## M67 (NGC 2682): Open Star Cluster in Cancer

DISTANCE FROM EARTH: 2,500 light-years
FINDING FACTOR: **
"WOW!" FACTOR:    Binoculars: **        Small telescopes (3″ to 5″): ***
                 Medium telescopes (6″ to 8″): ***

### Where to Look

M67 is not quite as easy to find as better-known M44, but it is still bright enough to be seen through finderscopes and binoculars. Begin by finding the trapezoidal head of Hydra, just peaking over the southern (bottom) edge of Spring Sky Window 1. Use the top three stars of Hydra's head as an arrow to point the way to M67, which sits just to the west of a 4th-magnitude star in Cancer named Acubens.

### What You'll See

**Through Binoculars.** M67 looks like a misty glow set between Acubens and two fainter stars to the west. Since the stars in M67 shine no brighter than 10th magnitude, they are too faint to resolve through most binoculars. One or two very faint points might be seen using averted vision, however, through binoculars braced on a tripod.

**Through a Telescope.** It's a shame that many observers overlook M67 in favor of the Beehive Cluster, since it is such a pleasant open cluster in its own right. In fact, once they compare the two, most people agree that M67 (Figure 6.5) is actually much prettier, since it is smaller and can therefore fit into an eyepiece field all at once.

When viewed through a 4-inch (10-cm) or smaller telescope at low power, the stars of M67 may appear to be engulfed in the gentle glow of nebulosity. This is just an illusion of aperture and magnification. By upping the telescope's power to 50× or more, the "cloud" is dispersed, leaving in its place myriad dim suns. Take a careful look and you'll see that many glow with subtle tints of yellow, orange, and red. These colors tell astronomers that M67 is composed of "mature" stars. Astronomers believe that M67 is about 3 to 4 billion years old, making it one of the oldest open clusters in the Milky Way and nearly as old as our solar system.

### M67

M67 contains five hundred stars scattered across half a degree. The age of M67 is based on where its stars are in their evolutionary process. To determine this, professional astronomers pass the light from each star through special attachments on their telescopes called spectrographs. These break the starlight into the spectrum, allowing astronomers to determine, among other things, each star's composition, temperature, and spectral class. Many of the stars in M67 are red giants, spectral type K and M, an advanced stage in a star's evolutionary life. The more red giants in a cluster, the older the cluster itself. Normally, open star clusters are found along the plane of our galaxy, the Milky Way. Gravitational influence from other stars usually causes star clusters to disperse in a matter of a few million years. But M67 is uniquely situated far from the Milky Way's plane, away from the influence of other stars. As a result, it has outlasted nearly every other open cluster in the Milky Way.

**Figure 6.5** *M67, open star cluster in Cancer. South is to the upper right. Photograph by George Viscome through a 14.5-inch f/6 reflector.*

## Iota Cancri: Binary Star in Cancer

DISTANCE FROM EARTH: 333 light-years
FINDING FACTOR: **
"WOW!" FACTOR:    Binoculars: not resolvable        Small telescopes (3″ to 5″): ***
                 Medium telescopes (6″ to 8″): ***

### Where to Look

Often drawn as the northern claw of the Crab, Iota Cancri can be seen as a dim point of light to the naked eye. If light pollution is too obtrusive to make it out, you can find Iota by aiming halfway along a line drawn between the star Castor in Gemini and Epsilon Leonis, the star at the tip of Leo's sickle.

### What You'll See

**Through Binoculars.** While visible through binoculars, Iota masks its split personality well. The two stellar components appear so close to one another that they are just barely resolvable through 10× binoculars. Even then, your best chance at seeing the two closely set stars is by bracing your binoculars against a tree, maybe a table, or ideally on a tripod or other mounting.

**Through a Telescope.** One look at Iota through any backyard telescope that is 2.4 inches (6 cm) and larger and you are sure to say "wow!" Iota is a beautiful binary that glistens in just about any instrument. The colors of the system's two stars are striking, the golden primary sun teamed with a bluish companion set to its northwest. It's sure to become a seasonal favorite.

## Spring Sky Window 2

The familiar pattern of the Big Dipper's bowl frames Spring Sky Window 2 (Figure 6.6). Visible throughout the year from much of the Northern Hemisphere, this section of sky offers up some very interesting galaxies and a lone planetary nebula to be enjoyed through backyard telescopes.

### M81: Spiral Galaxy in Ursa Major

NICKNAME: Bode's Nebula
DISTANCE FROM EARTH: 12 million light-years
FINDING FACTOR: **
"WOW!" FACTOR:   Binoculars: ***   Small telescopes (3″ to 5″): ****
                Medium telescopes (6″ to 8″): ****

#### Where to Look

M81 lies at the end of a long line extending to the northwest from Phecda through Dubhe, both in the bowl of the Big Dipper. The galaxy and Phecda are both about equal distances from Dubhe but set directly opposite each other. Looking through your finderscope, star hop (more like star bound) from Phecda to Dubhe, then on toward M81. You will pass a sparse arc of faint stars along the way, then come to a small right triangle. The star marking the right angle itself is cataloged as 24 Ursae Majoris. The triangle lies just to the northwest of M81 (and M82, discussed later).

#### What You'll See

**Through Binoculars.** Although rather dim, M81's oval form can be distinguished through binoculars as small as 7 × 35. Giant binoculars (that is, 10 × 70 and greater) heighten the contrast between the galaxy's bright central nucleus and the dimmer surrounding halo of the spiral arms. The spiral arms themselves are not nearly bright enough to be seen through binoculars.

**Through a Telescope.** Trying to see M81's spiral arms through even large backyard telescopes is daunting as well. Instead, most telescopes show the galaxy as a bright, oval glow surrounding a prominent central core, as seen in Figure 6.7. The galaxy appears oriented northwest-southeast, measuring about twice as long as it does wide. On exceptional nights, an 8-inch (20-cm) instrument just might show some vague irregularities across the spiral halo, but all in all, M81 shows less internal detail than many other bright spiral galaxies.

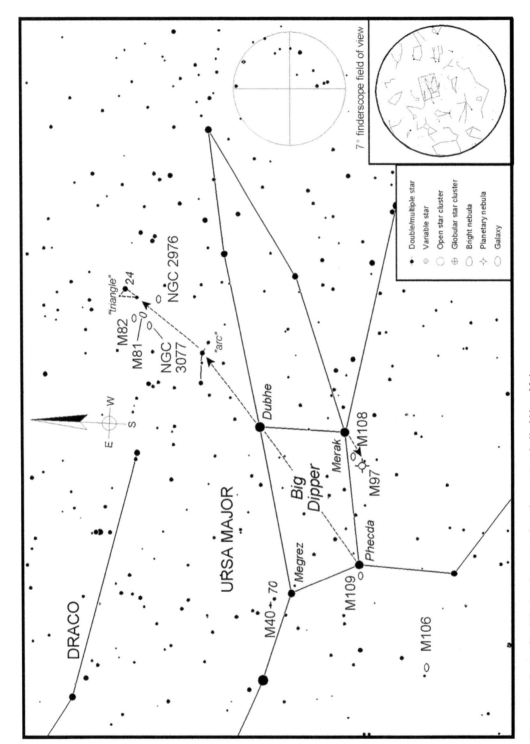

**Figure 6.6** *Spring Sky Window 2, centered on the western half of Ursa Major.*

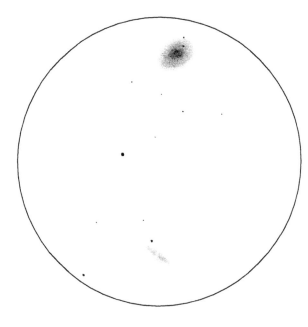

**Figure 6.7** *The galaxies M81 (top) and M82 (bottom) in Ursa Major, as drawn through the author's 8-inch reflector at 55×. South is up.*

### M81 Galaxy Group

M81 and M82 form one of the most intriguing pairs of galaxies found anywhere in the sky. Were we standing on a planet orbiting a star in one of the galaxies, we would have an amazing view of the other, since they lie within about 100,000 light-years of each other. M81 measures about 36,000 light-years from edge to edge, while M82 is comparatively small, only 16,000 light-years at its widest. M81, the brighter of the two by more than a full magnitude, was discovered by Johan Bode in 1774 and is a classic example of an Sb spiral (Figure 1.1). Color photographs show a yellowish tint to M81's core, the result of the proliferation of older stars that astronomers call Population II stars. The spiral arms appear bluish white, the result of younger, hotter Population I stars.

Also discovered by Bode in 1774, M82 is an irregular galaxy, a strikingly different kind of beast. Very long exposure photographs reveal a pair of huge nebulous plumes extending from the dark rift that cuts through the galaxy's core. Many feel these are clouds of material erupting from within the galactic nucleus. The detection of intense radio noise seems to further advance the theory of an exploding M82. But what is causing such stupendous destruction, detonating star after star? One thought is that M82 survived a collision with M81 in the far-distant past, which resulted in massive starburst activity.

Two other galaxies that are within reach of most amateur instruments also belong to the M81 Galaxy Group. Southwest of M81 lies NGC 2976, an oval-shaped spiral characterized by wide arms in photographs. Unfortunately, NGC 2976 is so dim that a 6-inch (15-cm) or larger telescope is needed just to see it. Then there is NGC 3077, set to the east-southeast of M81, near an 8th-magnitude star. A 4-inch (10-cm) or larger telescope will show its dim disk as a soft glow.

## M82: Irregular Galaxy in Ursa Major

NICKNAME: Cigar Galaxy
DISTANCE FROM EARTH: 12 million light-years
FINDING FACTOR:  **
"WOW!" FACTOR:    Binoculars: ***    Small telescopes (3" to 5"): ****
Medium telescopes (6" to 8"): ****

### Where to Look

M82 lies in the same low-power eyepiece field as M81. Look for its thin disk just to M81's northeast. M82 is about a magnitude fainter, but because of its unique elongated shape, it is fairly easy to spot once M81 is in view.

### What You'll See

**Through Binoculars.** The cigar shape of M82 can be seen through 7 × 35 and larger binoculars, as long as the night is dark and free from high levels of light pollution. Unlike M81, which shows a brighter core, M82 appears pretty much uniform in brightness from end to end.

**Through a Telescope.** Closer inspection of the galaxy through 6-inch (15-cm) and larger telescopes might show a jagged rift of darkness that rips across its center. With 10-inch (25-cm) and larger telescopes, other dark filaments can be spotted throughout M82, with the more conspicuous seen toward the galaxy's eastern edge.

## M97: Planetary Nebula in Ursa Major

NICKNAME: Owl Nebula
DISTANCE FROM EARTH: 2,600 light-years
FINDING FACTOR:  ***
"WOW!" FACTOR:    Binoculars: *    Small telescopes (3" to 5"): **
Medium telescopes (6" to 8"): ***

### Where to Look

M97 is easy to locate, but paradoxically, it is hard to see because of its dimness. To get there, aim your telescope toward Merak, the southwestern (lower right) star in the bowl of the Big Dipper. Looking through your finderscope, take aim at a dim star that lies about a degree to the east-southeast. Switch to your telescope, center on that star, then look just to its east for the Owl's dim, grayish disk.

### What You'll See

**Through Binoculars.** While the Owl Nebula is technically visible through binoculars (I've seen it through 10 × 50s), it really isn't a suitable object for study. If you want, give it a go and see if you can spot it. But don't be surprised if you can't find it, even with giant binoculars.

**Through a Telescope.** Depending on your telescope's aperture, the Owl Nebula may appear as little more than a faint glow or it may show some of the unique personality that has been recorded on photographs. A 6-inch (15-cm) telescope, for instance, may just begin to add a little character to the Owl by showing hints of two dark, round areas offset to either side of center. These are the famous "eyes" that gave rise to M97's nickname. Larger instruments add further definition to the eyes, but you'll need a very large telescope to reveal the nebula's 16th-magnitude central star.

The Owl is perched less than one degree to the southeast of our next target, the galaxy M108. In fact, a low-power eyepiece might just be able to squeeze both into the same field of view.

### Owl Nebula

The Owl was discovered by Pierre Mechain in 1781 and first viewed by Messier later that same year. Although it may look like a simple disk, the Owl Nebula is actually a complex object. The Owl is believed to be shaped like a donut. We are viewing it slightly off center, so the "eyes" of the Owl are actually the comparatively empty doughnut holes. The doughnut is also surrounded by a dimmer shell of expanding material.

Astronomers constantly debate the distance and diameter of the Owl. Some place it as close as 1,300 light-years from Earth, while others say that it is as far as 12,000 light-years away. Assuming that it is 2,600 light-years from us, the Owl is about 2 light-years across.

## M108: Spiral Galaxy in Ursa Major

DISTANCE FROM EARTH: 45 million light-years

FINDING FACTOR: ***

"WOW!" FACTOR:     Binoculars: *      Small telescopes (3″ to 5″): **
                  Medium telescopes (6″ to 8″): **

### Where to Look

M108 is located about 1.5 degrees to the southeast of Merak, the southern-most of the two pointer stars in the bowl of the Big Dipper. Look for it about halfway between Merak and M97, just described above.

### What You'll See

**Through Binoculars.** While M108 is technically bright enough to be seen through binoculars, it won't exactly "wow" you. Even peering through giant binoculars, you will see a pencil-thin smudge of grayish light.

**Through a Telescope.** Since M108 is tilted nearly edge-on from our earthly vantage point, its 10th-magnitude disk resembles a nebulous cigar nestled among an attractive field of stars. Although there is no evidence of spiral arms or a characteristic central bulge, M108 can reveal some faint, mottled surface detail if you really concentrate on it.

## M40: Double Star in Ursa Major

DISTANCE FROM EARTH: 510 light-years

FINDING FACTOR: **

"WOW!" FACTOR:    Binoculars: not resolvable    Small telescopes (3″ to 5″): *
Medium telescopes (6″ to 8″): *

### Where to Look

Aim toward Megrez, the star that joins the bowl of the Big Dipper with the handle. Viewing through your finderscope, look just to the northeast for a 6th-magnitude star cataloged as 70 Ursae Majoris. Switching to your telescope, center 70 in your lowest-power eyepiece, then look toward one edge of the field for a close-set pair of dim stars. That's M40.

### What You'll See

**Through Binoculars.** Sorry, but the two stars that form this double star are just too close and too faint to make much of an impression through binoculars.

**Through a Telescope.** Even through a telescope, M40 may win the award as the most disappointing Messier object. All you will find is a pair of white stars shining at magnitudes 9.0 and 9.3. At least the stars are easy to resolve, even at low power. Try a moderate magnification for the best view of this pair of white suns.

### Mistaken Identification

M40 was cataloged by mistake! The trouble began in 1660, when Hevelius recorded a "nebula above the back" of Ursa Major. Try as he might to duplicate the observation a century later, Messier could only find "two stars, very close together and of equal brightness, about 9th magnitude . . . it is presumed that Hevelius mistook these two stars for a nebula." Nevertheless Messier added the double star as his catalog's fortieth entry in 1764. In 1863, Hevelius's "nebula" was again discovered, this time by Friedrich August Theodor Winnecke at Pulkovo Observatory in St. Petersburg, Russia, who subsequently included it as the fourth listing in his double star inventory, entitled Doppelsternmessungen (Double Star Measurements). As a result, M40 is often cross-listed as Winnecke 4 in deep-sky catalogs.

## M109: Spiral Galaxy in Ursa Major

DISTANCE FROM EARTH: 55 million light-years

FINDING FACTOR: ****

"WOW!" FACTOR:    Binoculars: not resolvable    Small telescopes (3″ to 5″): **
Medium telescopes (6″ to 8″): **

### Where to Look

Locating M109 is easy, but seeing it is another matter. Aim your finderscope toward Phecda, the star that marks one of the bottom corners of the Big Dipper's bowl. Switch to your telescope, center Phecda in the field, then shift just a little to the southeast (maybe half a field). Look for a pair of 9th-magnitude stars (not shown on the chart, but visible through your telescope) that form a very narrow isosceles triangle with Phecda. M109 lies about halfway between those two faint stars. Have patience. It may take some time to pick out its faint glow.

### What You'll See

**Through Binoculars.** M109 is not bright enough to be visible through most common binoculars, although it may just break the visibility barrier in 70-mm and larger giants. Through these, it appears as little more than the dimmest smudge of light.

**Through a Telescope.** M109 can be a very challenging object to spot initially because of its low surface brightness. Don't get discouraged, it's there somewhere. Using averted vision, look for a very dim, oval blur with a slightly brighter core. The best view will probably be through a medium-power eyepiece (in the range of about 100×), although that depends on your telescope and local sky conditions. If you are using a wide-field eyepiece, such as a Nagler or Meade Ultra-Wide, be sure to move Phecda out of view, as its glare will easily overpower the feeble light from the galaxy.

## Spring Sky Window 3

The eastern half of Ursa Major, marked by the familiar curve of the Big Dipper's handle as well as portions of the faint constellation Canes Venatici (the Hunting Dogs), are surveyed in Spring Sky Window 3 (Figure 6.8). Like the previous Sky Window, much of this region remains visible throughout the year for most of the Northern Hemisphere. There's something for everybody here, ranging from a wide double star visible through binoculars to challenging galaxies.

### Alcor and Mizar: Double Star in Ursa Major

DISTANCE FROM EARTH: Mizar: 78 light-years, Alcor: 81 light-years
FINDING FACTOR:  *
"WOW!" FACTOR:     Binoculars: **     Small telescopes (3″ to 5″): ***
                  Medium telescopes (6″ to 8″): ***

### Where to Look

Alcor and Mizar are easily found, as they collectively mark the bend in the handle of the Big Dipper.

### What You'll See

**Through Binoculars.** Mizar teams with Alcor to form the most famous naked-eye double star in the entire sky. Separated by nearly 12' of arc, the pair gleams through even the smallest binoculars. Mizar is the brighter of the two, at magnitude 2.2, while Alcor is dimmer at magnitude 4.0. You might also see

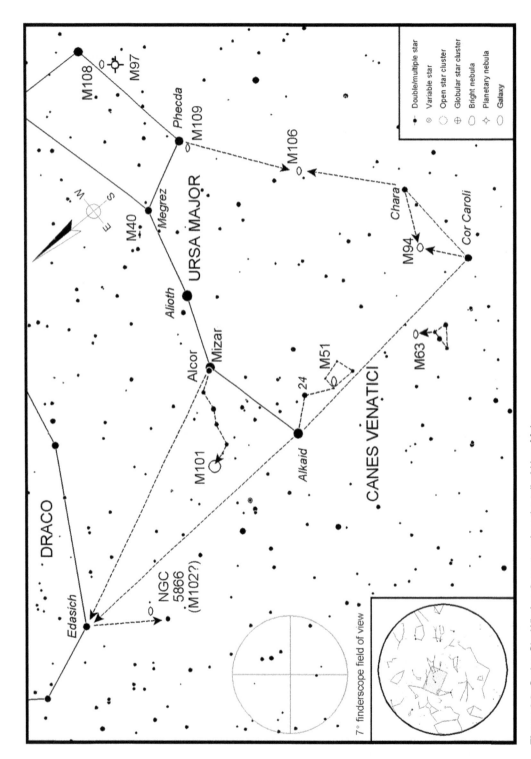

**Figure 6.8** *Spring Sky Window 3, centered on the tail of Ursa Major.*

an 8th-magnitude field star through binoculars that joins Alcor and Mizar to form a flattened triangular pattern.

**Through a Telescope.** A surprise awaits observers looking through telescopes. It turns out that Mizar is itself a close-set binary star, as you can see through even a 2-inch (5-cm) telescope. In fact, Mizar was the first double to be discovered telescopically when the Italian astronomer Giambattista Riccioli noticed its duality in 1650. Both suns appear white.

---

### *Alcor and Mizar*

Separated in space by 3 light-years, Alcor and Mizar are not true physical companions. The distance between them is too far for them to form a true binary system, but both stars do appear to be traveling through space together, along with several other stars in the Big Dipper. It turns out that five of the Big Dipper stars (Merak, Phecda, Megrez, Alioth, and Mizar), as well as approximately a hundred others, form a loose star cluster that has been dubbed the Ursa Major Moving Cluster.

Spectroscopic observations of Mizar show that both Mizar A and Mizar B are themselves binary stars, which makes Mizar a quadruple star system. Unfortunately, their companion suns are too close to be observed directly as separate stars even by the largest telescopes.

Some sources claim that Alcor is also a spectroscopic binary star, although other authorities now question this conclusion.

That third, dim field star forming a broad triangle with Alcor and Mizar holds an interesting footnote in astronomical history. In 1722, the German mathematician Johann Liebknecht thought he saw the star move against the background from one night to the next. He concluded that it was not a star at all but rather a new planet. In his excitement, he christened it Sidus Ludoviciana ("Ludwig's Star") after Ludwig V, then the king of Germany. It subsequently became apparent that Liebknecht was mistaken, but the star is still called Sidus Ludoviciana.

---

## M51: Spiral Galaxy in Ursa Major

NICKNAME: Whirlpool Galaxy
DISTANCE FROM EARTH: 31 million light-years
FINDING FACTOR: **
"WOW!" FACTOR:    Binoculars: **        Small telescopes (3″ to 5″): ****
                 Medium telescopes (6″ to 8″): ****

### Where to Look

Begin at Alkaid, the end star in the handle of the Big Dipper. Hop to the star 24 Canum Venaticorum, a 4th-magnitude point just over the border in the constellation Canes Venatici. From here, visualize an isosceles triangle formed by Alkaid at the northern corner, 24 Canum as the center point, and a third point an equal distance away from the star 24 Canum as is Alkaid. If you aim your binoculars or telescope's finderscope at this area and look carefully, you should notice a rectangle of four faint stars. M51 lies just inside the rectangle's northeast corner. The galaxy itself just might be visible through a 6 × 30 finderscope from reasonably dark skies.

### What You'll See

**Through Binoculars.** Steadily supported 7× and larger binoculars show M51 as a round, dim glow inside that rectangle mentioned above. Don't just grab a quick glance, then go. Instead, sit back and study the object. If you are viewing under dark skies, you might notice that M51 is a little lopsided—that there is a lump protruding on the north side of the galaxy.

That lump actually turns out to be another galaxy! Since Messier missed it, this companion galaxy is usually referred to by its number in the New General Catalog, NGC 5195. NGC 5195 glows weakly at about 10th magnitude, and as such, proves challenging in anything less than 50-mm binoculars.

**Through a Telescope.** M51 (Figure 6.9), a classic face-on spiral, arguably shows more subtle detail than any other galaxy visible from midnorthern latitudes. A quick glance through a 3- or 4-inch (7.5- or 10-cm) telescope will immediately show its slight oval disk and bright central nucleus. NGC 5195 is also apparent, as is an arm reaching out from M51, seemingly to embrace it.

Given dark, clear skies, high-quality 6- to 8-inch (15- to 20-cm) instruments can readily show the famous pinwheel-like arms of M51 wrapping around its bright central core. Although the spiral arms are rather subtle and may require a concentrated effort using averted vision, they help rank M51 as one of the sky's true showpieces. If your telescope isn't quite large enough to show the spiral arms, try to get a view of it through a larger instrument, such as at a star party hosted by a local astronomy club. It will be worth the trip.

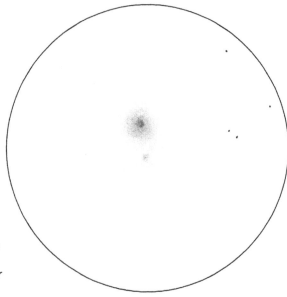

**Figure 6.9** *M51, the Whirlpool Galaxy in Canes Venatici, as drawn through the author's 8-inch reflector at 120×. South is up.*

### *My Favorite Galaxy: M51*

M51 is a textbook example of an Sc spiral galaxy. William Parsons, the third earl of Rosse, was the first to detect its pinwheel-like structure in 1845 while viewing through his mammoth 72-inch (1.8-meter) reflector from Birr, Ireland, then the world's largest telescope. Astronomers of the day thought that such "spiral nebulae" were solar systems in formation. It was not until the 1920s, when Edwin Hubble conceived the true galactic organization of the universe, that they were recognized as remote galaxies.

Although Messier himself discovered M51 on October 13, 1773, he evidently missed NGC 5195. NGC 5195 was discovered by Pierre Mechain in 1781 and subsequently noted by Messier as "double, each has a bright center . . . the two 'atmospheres' touch each other, the one is even fainter than the other." Messier failed to recognize that he was seeing two separate objects, so he gave the number M51 to both galaxies together. Today, however, the *M* designation is usually used to mean the spiral galaxy, while the NGC number is used for its companion.

NGC 5195 is an irregular galaxy, appearing as a fuzzy, not-quite-round glow seemingly linked to M51 by a luminous bridge of nebulosity. Indeed, the gravitational pull of NGC 5195 appears to have pulled this material from M51. At the same time, this shuffling of interstellar material has also triggered an increase in star formation in some regions of M51.

M51, believed to contain at least 100 billion stars, measures at least 50,000 light-years in diameter.

## M101: Spiral Galaxy in Ursa Major

NICKNAME: Pinwheel Galaxy
DISTANCE FROM EARTH: 27 million light-years
FINDING FACTOR: ****
"WOW!" FACTOR:    Binoculars: *        Small telescopes (3″ to 5″): **
                 Medium telescopes (6″ to 8″): ***

### Where to Look

M101 is located off the handle of the Big Dipper. To find M101, hop off from Alcor and Mizar, located at the crook in the handle, and head more or less due east along a path of five dim stars. At the fifth star, hop a little to the northeast to spot M101.

Be forewarned, however, that M101 is one of those objects that even seasoned observers have trouble locating. Although M101 is rated at 8th magnitude, which is actually quite bright for a galaxy, that value indicates its equivalent stellar magnitude if the galaxy could somehow be squeezed down to a point. Since M101 appears considerably larger than a point, its magnitude has been spread out to match its diameter. Therefore, its "surface brightness," or brightness per unit area, is very low. That's the problem. For tips on dealing with low surface brightness, see the discussion of M33 in chapter 8.

### What You'll See

**Through Binoculars.** This is a tough one through small telescopes, let alone binoculars, but it is possible. Observers stand the best chance of spotting M101's faint galactic nucleus by using averted vision, provided the binoculars are firmly braced against a steady support. If you are convinced your aim is correct but still can't find the galaxy, try jiggling the glasses slightly. Sometimes objects that are just barely detectable can be spotted if the field of view is shaken ever so slightly.

**Through a Telescope.** Although M101 will probably impress you as little more than a featureless glow at first, its beauty can really blossom with closer inspection, that is, if you have a large enough telescope aperture, great optical quality, and excellent sky transparency.

Under exceptional skies, an 8-inch (20-cm) telescope at about 100× can show a subtle hint of the spiral pattern seen so prominently in Figure 6.10. The brightest spiral arm curves away from the core to the north and extends toward the east. Following the arm, look for several bright patches that mark either star clouds or giant patches of nebulosity.

## NGC 5866 (M102?): Lenticular Galaxy in Draco

NICKNAME: Spindle Galaxy
DISTANCE FROM EARTH: 40 million light-years
FINDING FACTOR: ****
"WOW!" FACTOR:    Binoculars: not resolvable      Small telescopes (3″ to 5″): **
                 Medium telescopes (6″ to 8″): **

### Where to Look

Unfortunately, NGC 5866 is out in the middle of nowhere, so trying to pinpoint exactly where to aim your finderscope is difficult. First, find the star Edasich in the tail of Draco. It might be helpful to imagine an isosceles triangle with Mizar and Alkaid in the handle of the Big Dipper as the two endpoints of the base of the triangle and Edasich as the apex. Once you've found Edasich, position it on the northeastern edge of your finder's field of view and look for a 5th-magnitude star toward the southwestern edge. NGC 5866 lies just to the east-northeast of that second star.

### What You'll See

**Through Binoculars.** Shining only at 10th magnitude, NGC 5866 is beyond the reach of all but the largest binoculars.

**Through a Telescope.** NGC 5866 appears highly elongated, stretching nearly three times as long as it appears wide. The galaxy's edge-on orientation, as seen from our perspective, causes the ends of its disk to taper down to thin tips, disappearing into the depths of intergalactic space. Astronomers call this a *lenticular* galaxy because its profile is similar to that of a double-convex lens.

**Figure 6.10** *M101, the Pinwheel Galaxy in Ursa Major. South is up in this photograph by George Viscome through a 14.5-inch f/6 reflector.*

This unusual shape also gives NGC 5866 its nickname the Spindle. A thin line of dark nebulosity slices across the edge of the galactic plane, although spotting it is a tough assignment.

### The Spindle Galaxy

NGC 5866 is classified as a lenticular, or type S0, galaxy. S0 galaxies share many of the features found in elliptical galaxies, including a lack of nebulosity, but they also share many features of spiral galaxies, such as a central nuclear bulge and a noticeable disk, but with one important difference. There is no sign of spiral arms. Instead their spiral structure has spread out evenly across the galaxy's disk, which measures about 60,000 light-years in diameter.

Much controversy has surrounded NGC 5866 since the 1780s, with some authorities cross-listing it as M102 and others not. Did Messier actually observe this galaxy? Apparently, Pierre Mechain, credited as "discovering" M102, didn't think so. There is other evidence, however, that Messier did indeed observe NGC 5866 independently, and so it can rightfully be called M102.

## M63: Spiral Galaxy in Canes Venatici

NICKNAME: Sunflower Galaxy
DISTANCE FROM EARTH: 37 million light-years
FINDING FACTOR: **
"WOW!" FACTOR:    Binoculars: **      Small telescopes (3″ to 5″): ***
                 Medium telescopes (6″ to 8″): ***

### Where to Look

M63 is found about two-thirds of the way between the stars Alkaid, the end star in the handle of the Big Dipper, and Cor Caroli. As you slide from one star to the other while looking through your binoculars or finderscope, keep an eye out for a distinctive pattern of four stars in the shape of a triangle. M63 lies just to its north.

### What You'll See

**Through Binoculars.** M63 can be seen through 10 × 50 and larger binoculars, although it will be just barely visible from light-polluted suburbs. Look for an ill-defined smudge of light set close to an 8th-magnitude field star. Giant binoculars reveal M63 as distinctly cigar-shaped and without any bright central nucleus.

**Through a Telescope.** Nicknamed the Sunflower Galaxy, M63 looks like a silvery sliver of light brightening slowly at first, then rapidly toward an inner core. An 8-inch (20-cm) telescope adds a brighter nucleus. M63 bears magnification well, with the best views coming between 150× and 200×.

## M94: Spiral Galaxy in Canes Venatici

DISTANCE FROM EARTH: 14 million light-years
FINDING FACTOR: **
"WOW!" FACTOR:    Binoculars: **      Small telescopes (3″ to 5″): **
                 Medium telescopes (6″ to 8″): ***

### Where to Look

Spiral galaxy M94 lies a little northeast of the halfway point between the stars Cor Caroli and Chara in Canes Venatici, which you will find under the curve of the Big Dipper's handle. Cor Caroli shines at 3rd magnitude, while Chara is about a magnitude dimmer, so both should be visible to the eye alone. Try to visualize the triangle shown on Figure 6.8 and aim your telescope toward the triangle's third point.

### What You'll See

**Through Binoculars.** M94 is bright enough to be visible through binoculars as small as 7 × 35, even through moderately light-polluted suburban skies. Look for a concentrated, nearly circular halo that appears almost starlike. In fact, the galaxy is so concentrated you might think that it is just another star at first glance. A careful check, however, will show that M94 looks a little fuzzy as compared to the other pinpoint stars in the neighborhood.

**Through a Telescope.** M94 still looks almost starlike through telescopes at lower magnifications, since we are only seeing the galaxy's concentrated nucleus. Its spiral arms, which are quite tightly wrapped in this case, are too faint to be detected easily through amateur telescopes.

### M106: Spiral Galaxy in Canes Venatici

DISTANCE FROM EARTH: 25 million light-years
FINDING FACTOR:  **
"WOW!" FACTOR:    Binoculars: **      Small telescopes (3″ to 5″): **
                  Medium telescopes (6″ to 8″): ***

#### Where to Look

M106 is found almost directly between the stars Phecda, in the bowl of the Big Dipper, and Chara in Canes Venatici. As shown on Figure 6.8, it lies just to the north of a 6th-magnitude field star, which should be visible through your finderscope.

#### What You'll See

**Through Binoculars.** M106 can be spotted through 50-mm binoculars as a faint smudge of grayish light next to that 6th-magnitude foreground star. Its oval shape should be clear in 10× and higher powered binoculars, as should its brighter central nucleus.

**Through a Telescope.** Through a 4- to 6-inch (10- to 15-cm) telescope, M106 shows a bright, oval disk that looks a little like an underinflated football. Eight-inch (20-cm) and larger instruments begin to add some mottling to the dimmer, outer edges of M106, the galaxy's spiral-arm halo. Try a medium magnification (about 100×) for the best view.

## Spring Sky Window 4

The conspicuous constellation of Leo the Lion with its bright star Regulus dominates Spring Sky Window 4 (Figure 6.11). Several prominent galaxies lie within the confines of this area, many of which are even visible, albeit faintly, through binoculars. The stars of Leo serve as convenient pointers toward each.

### M65: Spiral Galaxy in Leo
### M66: Spiral Galaxy in Leo
### NGC 3628: Spiral Galaxy in Leo

NICKNAME: Leo Triplet Group
DISTANCE FROM EARTH: 35 million light-years
FINDING FACTOR:  **
"WOW!" FACTOR:    Binoculars: **      Small telescopes (3″ to 5″): **
                  Medium telescopes (6″ to 8″):  **

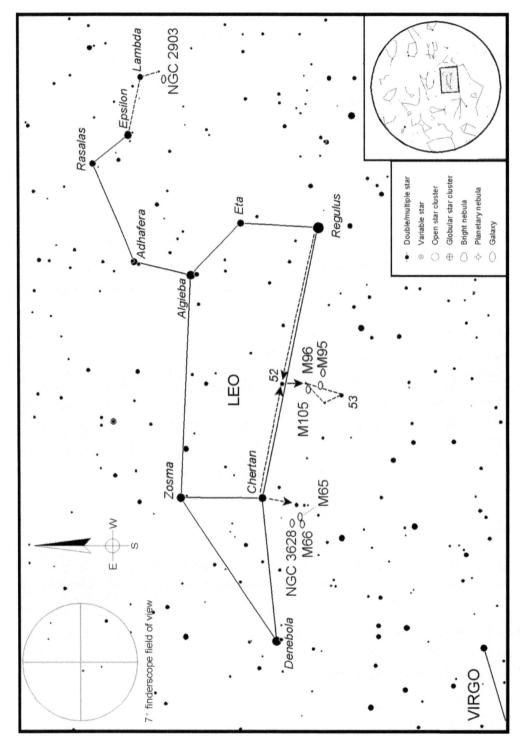

**Figure 6.11** *Spring Sky Window 4, framing the distinctive form of Leo the Lion.*

### Where to Look

To find this small group of galaxies, start at Leo's hind triangle. Extend a line from the star Zosma south to Chertan and continue southward, looking through your finderscope or binoculars to a line of three faint stars. All are 7th magnitude or brighter, and so should be visible through 6 × 30 and larger finders. Switching to your telescope and using your lowest-power eyepiece, shift the aim about one eyepiece field east to find M65 and M66. NGC 3628 is just to their north-northeast.

### What You'll See

**Through Binoculars.** Through 10× binoculars, M66, the brightest of the three, looks like an oval disk with a brighter center. Although slightly fainter, M65 is nearly identical in both visual appearance and form. NGC 3628 is probably too faint to be seen through most binoculars.

**Through a Telescope.** Once again, M66 appears to be the dominant galaxy, outshining its two neighbors. Since its spiral arms are rather faint, however, a 4- to 6-inch (10- to 15-cm) telescope will only reveal the galaxy's brighter, oval core. Eight- to 10-inch (20- to 25-cm) instruments add hints of many bright and dark patches throughout the galaxy's spiral-arm halo.

M65, found in the same medium-power field of view, appears longer, thinner, and a little fainter than M66. Although also classified as a spiral galaxy, M65 appears more edge-on to our line of sight.

The third galaxy here, NGC 3628, is larger and fainter than both M65 and M66. Its long, thin disk shines at about 10th magnitude and is quite extended, measuring ten times as long as it does wide in photographs. Like many edge-on spiral galaxies, NGC 3628 is bisected by a prominent dust lane that passes through the central core and along the outer edge of the spiral-arm disk.

## M95: Barred Spiral Galaxy in Leo
## M96: Spiral Galaxy in Leo
## M105: Elliptical Galaxy in Leo

NICKNAME: M96 Galaxy Group
DISTANCE FROM EARTH: 38 million light-years
FINDING FACTOR: **
"WOW!" FACTOR:    Binoculars: **        Small telescopes (3″ to 5″): **
                 Medium telescopes (6″ to 8″): **

### Where to Look

This trio of galaxies can be found about midway between the stars Regulus and Chertan, along the belly of Leo. Aim your finderscope or binoculars there and you'll come to a 5.5-magnitude star labeled 52 Leonis. Just to its south, you'll see a faint triangle of 6th- and 7th-magnitude stars, the brightest and southernmost labeled 53 Leonis. The target galaxies lie within or adjacent to that triangle.

### What You'll See

**Through Binoculars.** All three galaxies should be visible, though very faintly, through 50-mm binoculars. M96 will probably strike you as the brightest of the three, although it is easily confused with M95, just to its west. To tell one from the other, double-check Figure 6.11 and compare the galaxies' positions with the triangle of stars noted earlier. M105 is slightly fainter than the other two, located an equal distance to the northeast of M96 as M95 is to its west.

**Through a Telescope.** Through most amateur telescopes M96 looks like a nebulous glow surrounding a brighter, oval heart. Although fairly bright as galaxies go, M96 gives up little additional detail even through double-digit apertures. M95 lies just west of M96, almost squeezing into the same low-power eyepiece field. With an 8-inch (20-cm) telescope, it appears as a nearly circular patch of grayish light growing brighter toward its center. The core itself is oval and spotlighted by a sharp stellar nucleus. Lastly, M105 is also within the same low-power eyepiece field as M95 and M96. Visually, it appears smaller than either of its neighboring spirals and lies about halfway between them in terms of brightness.

## NGC 2903: Spiral Galaxy in Leo

DISTANCE FROM EARTH: 25 million light-years
FINDING FACTOR: ***
"WOW!" FACTOR:    Binoculars: *        Small telescopes (3″ to 5″): **
                 Medium telescopes (6″ to 8″): ***

### Where to Look

To find NGC 2903, start at the star Epsilon Leonis, which marks the western tip of Leo's backward question mark. Center it in your finderscope, then look just to its west, to the faint star Lambda Leonis. You should see an even fainter star to Lambda's southwest. Shift your finderscope toward that southwestern star, then switch to your telescope and your lowest-power eyepiece. NGC 2903 should look like a faint blur just to the east of that star.

### What You'll See

**Through Binoculars.** NGC 2903 is one of the brightest galaxies in the spring sky that does not belong to the Messier catalog. But even with that distinction, spotting its dim glimmer through handheld binoculars is definitely a challenge. It shines weakly at 9th magnitude, just at the edge of visibility in 50-mm binoculars when used under near-perfect sky conditions. Any degree of light pollution will probably render it invisible.

**Through a Telescope.** Three- to 6-inch (7.5- to 15-cm) telescopes reveal that 9th-magnitude NGC 2903 has a noticeably oval disk and a brighter central hub. Larger telescopes also show some knots and irregularities within the galaxy's disk, probably coinciding with concentrations of stars and nebulae.

## Spring Sky Window 5

The fifth Spring Sky Window (Figure 6.12) includes more objects in a smaller area than any other sky window in this book. Since it bridges the constellations Coma Berenices and Virgo, astronomers call this region the Coma-Virgo Realm of Galaxies. There are sixteen Messier objects within the Realm, as well as scores of other galaxies that are listed in the *New General Catalog* (NGC). Most of the non-Messier galaxies are very faint, although the few high-lighted here are visible through 3- and 4-inch (7.5- and 10-cm) telescopes.

The key to making your way through the Realm of Galaxies is to adopt a slow, patient approach. Take your time while hopping from one object to the next, following the directions offered below.

### The Coma-Virgo Galaxy Cluster

NICKNAME: The Realm of Galaxies
DISTANCE FROM EARTH: 60 million light-years
FINDING FACTOR: ***
"WOW!" FACTOR:    Binoculars: *        Small telescopes (3″ to 5″): **
                 Medium telescopes (6″ to 8″): ***

#### Where to Look

While some books treat this gaggle of galaxies as individual objects, I have always found it easier to attack them en masse, staging my assault in two waves: one entering the Realm from the west, the other from the east.

For the western attack, begin at the star Denebola in the triangle of Leo the Lion. From there, move one finderscope field (about 7°) due east, to a small triangular asterism (pattern) formed by four faint stars. The northern-most star can also be shared with a diamond pattern seen just to the north.

The eastern assault on the Realm begins from the star Vindemiatrix in Virgo. From there, follow a path about one finderscope field due west, where you will be greeted by a pair of suns, including 5th-magnitude Rho Virginis and a fainter 7th-magnitude star.

#### What You'll See

**Through Binoculars.** While it's possible to see just about all of the Messier galaxies in the Realm through 50-mm binoculars, be forewarned that these are not suitable targets for those new to observing. Instead I recommend that you polish your skills on other, more impressive Messier objects first, then return here after those others have been exhausted.

Of the galaxies in the Realm, M49 and M87 are the brightest and easiest to see through binoculars. Both stand out from the crowd, revealing round or slightly oval disks, even at only 7×. Others that are bright enough to show themselves through most common-size binoculars include M60, M84, M85, M86, M88, and M90. Use the instructions below for wading through the Realm. Before you try something as challenging as the Realm of Galaxies, be sure

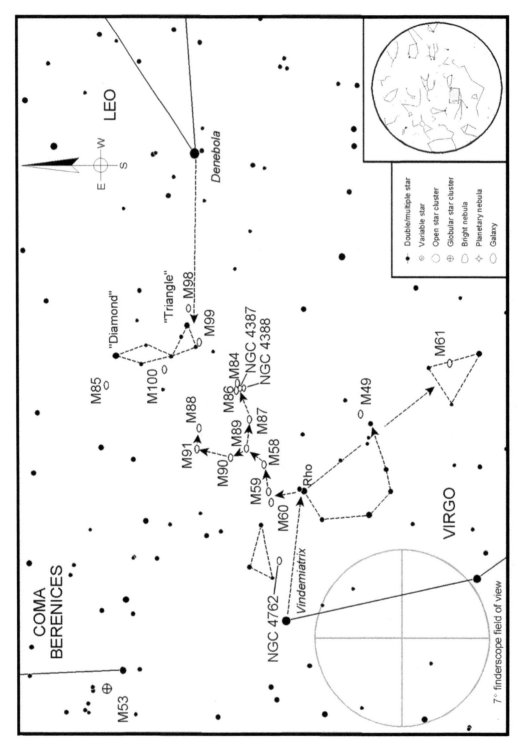

**Figure 6.12** *Spring Sky Window 5, showing a close-up view of the Coma-Virgo Realm of Galaxies.*

that your binoculars are mounted on some sort of tripod or other support. This advice includes image-stabilized binoculars. True, their technology will help calm jitters, but it is much easier to be able to go back and forth between binoculars and a chart without having to reaim the binoculars each time.

**Through a Telescope.** *Part 1: The Western Campaign.* From Denebola, slide eastward to that faint triangle of four stars shown on Spring Sky Map 5. M98, just west of that triangle, looks like a cigar-shaped shaft of grayish light in amateur telescopes. You may see the galaxy's prominent, circular core, but little additional detail is visible.

M99, sometimes nicknamed "The Pinwheel" for its spiral structure, lies just to the west of the triangle's southernmost star. M99 is bright enough to be seen through 3-inch (7.5-cm) telescopes on dark nights, while 6- and 8-inch (15- and 20-cm) telescopes are large enough to reveal one of the galaxy's spiral arms. (Don't be surprised if you don't see this arm, even from dark skies. It takes an experienced eye to make it out. Look for the arm curving toward the southwest, directly opposite that 7th-magnitude triangle star.)

M100, just east of the northernmost star in the triangle, displays a circular disk that becomes brighter toward its center. With an 8-inch (20-cm) telescope (possibly even a 6-inch [15-cm]), look for a subtle texture to the disk's outer circumference.

The western assault on the Realm ends with M85, found to the northeast of the diamond. Telescopes both small and large reveal it as an oval, silvery blur that suddenly brightens to an intense stellar core.

*Part 2: The Eastern Campaign.* The eastern path through the Realm is a bit more involved, with several twists and turns. While this book promotes star hopping for finding deep-sky objects, here we will have to rely on "galaxy hopping."

Just north of the halfway point between Vindemiatrix and Rho, look for a right triangle of three faint stars. Center your telescope on the star in the triangle's southeastern corner, then shift just slightly west to find NGC 4762, one of the Realm's brightest non-Messier members. NGC 4762, an unusual SB0 barred galaxy, looks like another cigar-shaped shaft of grayish light framed by a pair of 9th-magnitude stars to its east and west.

From Rho, follow a curve of stars to the south and west, to arrive at M49. The brightest member of the Realm, M49 is seen as a nearly circular ball of fuzz, accented with a brighter core. As mentioned below, M49 is one of the largest galaxies in the Realm, spanning an estimated 150,000 light-years.

Next, looking through your finderscope, draw an imaginary line southwestward from Rho, past several stars and head southwest to a triangle of three suns. About halfway along the triangle's western side, you'll find M61. M61 suffers from a low surface brightness, which causes many new observers to pass right by. A slow, concentrated scan is your best chance at seeing its subtle glow.

Go back to Rho, then head about an eyepiece field north to find M59 and M60. Both can probably just be squeezed into the same medium-power eyepiece field. M59, the eastern member of the team, is fairly obvious in all but the smallest backyard telescopes. M60 is both larger and brighter than M59, and in fact is one of the brightest galaxies in the Realm. Look for a round sphere of distant starlight marked by an obvious stellar core. M58 lies just east of M59 and M60. Most amateur instruments show M58 as an oval disk highlighted by a more intense central nucleus.

Continue the northwest-arcing curve that goes from M59 through M58 to get to M89. M89 shines at about 10th magnitude, which may make it a challenge through a 4-inch (10-cm) instrument, depending on light pollution. A 6-inch (15-cm) or larger shouldn't have much trouble picking it out. M89 looks almost perfectly circular and boasts a distinctive, stellar nucleus.

M89 lies in the same low-power field as M90, both apparently separated by an asterism of five 9th- and 10th-magnitude stars in a W-pattern reminiscent of Cassiopeia. M90 reminds me of the famous Andromeda Galaxy as seen through a finderscope or binoculars. Its spiral-arm halo is tilted almost due north-south, poised nearly edge-on to our line of sight.

## Coma-Virgo Realm of Galaxies

At an estimated 60 million light-years away, the Coma-Virgo Realm of Galaxies is the closest major cluster of galaxies to the Milky Way. While Messier cataloged sixteen galaxies here, that's not even the tip of this galactic iceberg! Studies show that as many as two thousand individual galaxies belong to this extended family. In fact, some authorities believe that our Milky Way and other galaxies in our immediate neighborhood (called the Local Group; see M31 for more details) are actually distant members as well. The combined mass of all the galaxies is so great that many are actually being pulled toward the cluster's center. Some even suggest that although the Milky Way is currently receding from the Virgo Cluster, the motion will eventually be reversed sometime in the unimaginably distant future. At that point, the Milky Way may be pulled into the cluster, perhaps to eventually merge with one or more other galaxies.

Galaxies M84, M86, and M87 lie near the center of the Coma-Virgo Cluster. As mentioned earlier, M87 is one of the most massive galaxies ever discovered, containing an estimated 2.7 trillion solar masses. Studies show that it measures about 120,000 light-years across, which is a little larger than the Milky Way. But while the Milky Way is a spiral, and therefore has most of its stars confined to a flat plane, M87 is classified as an E1 elliptical. As such, it takes up a much larger volume and is likely filled with more than 2 trillion stars!

There is more to M87 than just stars, however. In 1918, Heber Curtis of Lick Observatory discovered a strange jet of luminous material bursting from M87's core. Later observations revealed that the galaxy is also a strong emitter of radio and X-ray radiation, cross-cataloging it as the Virgo A radio source. In 1994, the Hubble Space Telescope found that the jet and emissions result from a massive gravitational collapse deep within the galactic core, likely powered by a black hole. A most unusual galaxy indeed!

M49 and M60 are also considered giants among the Realm's galaxies, with each holding more than a trillion stars.

Of the Messier galaxies in the Realm, only M91 is somewhat questionable. For years it was noted as "missing," based on Messier's notes, which ambiguously located it with respect to M58. In reality, he had actually noted its position away from M89, according to the amateur astronomer William C. Williams of Texas in 1958. Thanks to Williams's research, we know that M91 is galaxy NGC 4548.

Continue to arc toward the north-northeast to arrive at M91. A barred spiral, this galaxy appears fairly bright and large in 6- to 8-inch (15- to 20-cm) telescopes, although its shape seems vague at first. Careful examination shows a brighter, central core extended toward the northeast-southwest, lining up in the same direction as the galaxy's bar.

Hop a little less than a degree to the east-southeast from M91 to arrive at M88. M88, set just southeast of a 9th-magnitude field star, displays a bright nucleus that appears offset within its spiral-arm halo. Vague hints of spiral structure are visible, albeit with great difficulty, through telescopes as small as 8 inches (20 cm) in aperture. As with most of the Realm galaxies detailed here, use a low-power eyepiece to locate the target, then switch to a moderate-power eyepiece, producing an exit pupil between 2 mm and 3 mm, for the best views. (Remember, exit pupil is calculated by dividing the magnification by the telescope's focal ratio, or f/ number.)

Now, retrace your steps along the arc back to M89, then continue toward the west to M87. Amateur telescopes reveal the galaxy's bright core engulfed in a fainter, round mist that fades slowly toward the edge. Both M49 and M87 are considered "super galaxies," two of the largest galaxies ever found.

We wrap up the Realm by heading east-northeast just a bit to M84 and M86, the bright pair of galaxies in Figure 6.13, attended by several fainter systems that lie right at the heart of the Realm. Only a third of a degree separates M84 and M86 in our sky, so they easily fit into the same eyepiece field. Both feature bright central cores enveloped in dim halos made up of billions of stars that are so faint, we can only see their collective glow. A casual observer might see these 9th-magnitude galaxies as identical, but a closer look shows M84 to be the smaller and brighter of the pair, while M86 is slightly oval.

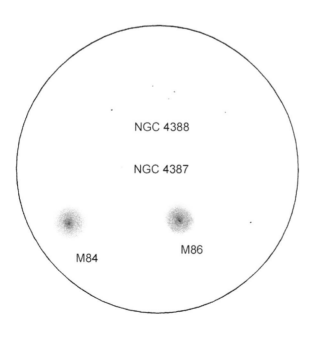

NGC 4388

NGC 4387

M84

M86

**Figure 6.13** *M84 and M86, two of the brightest galaxies in the Realm of Galaxies. Also note NGC 4387 and NGC 4388. Drawing based on the view through the author's 8-inch reflector at 55×. South is up.*

Several faint galaxies within the grasp of a good 8-inch (20-cm) telescope surround M84 and M86. Just 11' to their south is NGC 4387, a tiny elliptical, while NGC 4388, an edge-on spiral, is found a little farther south still. To many observers these four galaxies seem to form the shape of a giant intergalactic happy face! M84 and M86 mark the eyes, NGC 4387 the nose, and NGC 4388 the thin mouth.

## Spring Sky Window 6

Scraping the southern horizon, Spring Sky Window 6 (Figure 6.14) includes southernmost Virgo with its bright star Spica, as well as the dim constellation Corvus the Crow and the tail end of Hydra. Only three deep-sky objects await us here, although they each possess unique characteristics that make them stand out among the many objects in the spring sky.

### M104: Spiral Galaxy in Virgo

NICKNAME: Sombrero Galaxy
DISTANCE FROM EARTH: 50 million light-years
FINDING FACTOR: **
"WOW!" FACTOR:    Binoculars: **       Small telescopes (3″ to 5″): ***
                 Medium telescopes (6″ to 8″): ****

#### Where to Look

Begin by finding the constellation Corvus (remember, "curve to Corvus") southwest of the bright star Spica. Once there, set your sights on the star Gienah, which is often depicted as the Crow's head or beak. Just north of Gienah, you'll see a 6th-magnitude star, the first in a line of five 6th- and 7th-magnitude stars that extend to the northeast (only three of the stars are bright enough to show up on Figure 6.14, but all are connected with a dashed line for easy reference). Follow the line to its northern end, where you'll find a triangle of stars almost acting as an arrowhead. The arrow points to a small knot of faint stars. That small knot of stars seen between the arrowhead and M104 is a curious sight. This chance alignment of stars looks just like a triangle within a triangle. Although I've heard it given several names by "discoverers," the Texas amateur astronomer John Wagoner christened it the Stargate, which seems like a very good name to me, since it reminds me of a portal of sorts. Be sure to take a closer look when you have some time.

M104 lies just beyond the Stargate and should be visible through 50-mm finderscopes and binoculars.

#### What You'll See

**Through Binoculars.** Most binoculars show M104 as a tiny, oval disk that brightens rapidly toward the center. Giant binoculars may just add a hint of the band of dark nebulosity that slices across the edge of M104.

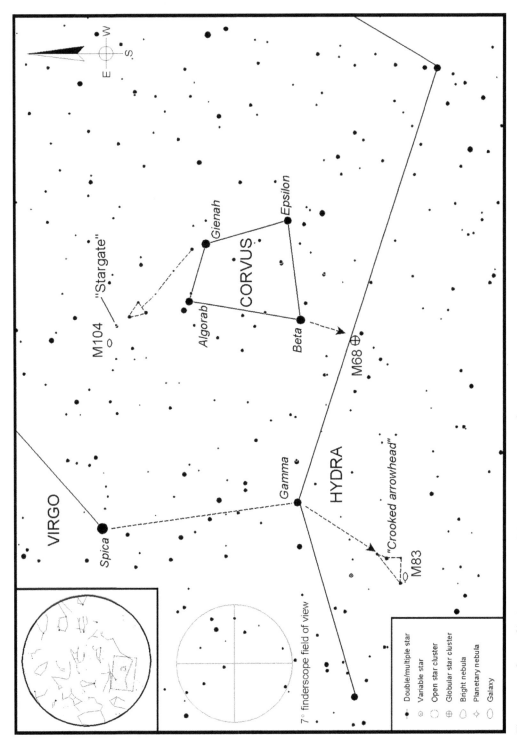

**Figure 6.14** *Spring Sky Window 6, highlighted by the bright star Spica.*

135

**Figure 6.15** *M104,
the Sombrero Galaxy,
as drawn through the
author's 8-inch reflector
at 120×. South is up.*

**Through a Telescope.** M104 (Figure 6.15) is a fun object to study! Through a telescope, you will see its protruding central core and broad, flattened spiral-arm rim, which is highlighted by a prominent dark lane encircling the galaxy's outer circumference. These features are why M104 is nicknamed the Sombrero Galaxy. When viewed through 4- to 6-inch (10- to 15-cm) telescopes, the Sombrero's dark lane may be seen cutting across the south side of the nucleus. The added light-gathering ability of an 8-inch (20-cm) instrument extends the lane fully across the galactic "brim" of M104.

## M68: Globular Cluster in Hydra

DISTANCE FROM EARTH: 33,000 light-years
FINDING FACTOR: ***
"WOW!" FACTOR:    Binoculars: *        Small telescopes (3″ to 5″): **
                 Medium telescopes (6″ to 8″): **

### Where to Look

To find M68, trace a line from Algorab in the constellation Corvus to Beta Corvi. Extend that line about half the distance farther to the south of Corvus, where you will come to a fairly bright field star (fairly bright through binoculars and finderscopes, that is). M68 lies just to its northeast.

When searching for M68, keep in mind that it is never very high above the horizon from midnorthern latitudes (for instance, it never climbs higher than about 15° above the horizon from latitude 40° north). Therefore a clear, unobstructed view to the south is a must. Even then, don't be surprised if some searching is required, especially if a thin layer of haze lies near the southern horizon.

## What You'll See

**Through Binoculars.** M68 is a relatively small, faint globular cluster through binoculars. At 8th magnitude, it remains only a dim ball of fluff.

**Through a Telescope.** Four-inch (10-cm) telescopes show a texture or "graininess" to the round globe of M68 but are unable to resolve any individual stars. By increasing the aperture to 6 inches (15 cm) or larger, a few stars right around the edges might be resolved, but even this is difficult.

## M83: Barred Spiral Galaxy in Hydra

NICKNAME: Southern Pinwheel Galaxy
DISTANCE FROM EARTH: 15 million light-years
FINDING FACTOR: ***
"WOW!" FACTOR:     Binoculars: **     Small telescopes (3″ to 5″): **
                  Medium telescopes (6″ to 8″): ***

### Where to Look

M83 is located even farther south in the sky than M68, so it is especially difficult to locate without a clear view. Start your hunt from the bright star Spica in Virgo. Head southward about two finder fields, to 3rd-magnitude Gamma Hydrae, which should certainly be visible with the naked eye as well. With your finderscope or binoculars, swing to the southeast, looking for a pattern of stars that resembles a crooked arrowhead. M83 lies just to the south of the arrowhead's point.

### What You'll See

**Through Binoculars.** Riding the Hydra-Centaurus border, M83 is bright enough to be seen through nearly all binoculars. Look for a bright stellar nucleus engulfed in the soft glow of the spiral arms. Thanks to the attractive star field created by that crooked arrowhead, M83 is sure to become one of your favorite seasonal binocular targets.

**Through a Telescope.** Were M83 better placed for observers in the Northern Hemisphere, it would undoubtedly be one of the season's most popular attractions. But like many of the finer things in life, its appeal may not be appreciated at first pass. Initially, M83 appears indistinct and shapeless, but with careful scrutiny through a 6-inch (15-cm) or larger telescope, an irregular texture can be detected with averted vision.

### M83

Discovered by Abbé Nicholas Louis de Lacaille during an expedition to South Africa in 1751–1752, M83 is a beautiful spiral galaxy with a rich structure. Long-exposure photographs show that its very well defined spiral arms are littered with clumps of red and blue, threaded together by many inky black dust lanes. The red knots are giant patches of glowing nebulosity where stars are being formed, while the blue regions are lit by the scattered light of young stars that have formed within the past several million years.

## Spring Sky Window 7

The final Spring Sky Window (Figure 6.16) includes the season's brightest star, Arcturus in Boötes the Herdsman, as well as the dim constellations Coma Berenices (Queen Berenices's Hair) and Canes Venatici (the Hunting Dogs). If you have sharp eyes and a dark sky, you may notice a large, hazy patch of grayish light in the sky just to the west of Arcturus. That's actually a star cluster, one of the most striking binocular objects of the season. Several galaxies, a pretty binary star, and two globular clusters, including the spring's finest, are also framed in this window.

### M3: Globular Cluster in Canes Venatici

DISTANCE FROM EARTH: 30,600 light-years

FINDING FACTOR: **

"WOW!" FACTOR:    Binoculars: ***    Small telescopes (3″ to 5″): ****
Medium telescopes (6″ to 8″): ****

#### Where to Look

The simplest way to find M3 is to aim your telescope just shy of the halfway point between the stars Arcturus in Boötes to Cor Caroli in Canes Venatici (found just south of the Big Dipper). M3 should be at or very near the center of your finderscope's field, looking like a fuzzy star just northeast of a faint sun.

#### What You'll See

**Through Binoculars.** Through 7× binoculars, M3 looks like a tiny puff of celestial cotton hanging near a 6th-magnitude field star.

**Through a Telescope.** A true superstar of the late spring sky, M3 is one of those happy objects that's a joy to view regardless of telescope aperture. Like Messier's own telescope, a 3-inch (7.5-cm) or smaller instrument will show a hazy smudge that gradually brightens toward the center. A 4-inch (10-cm) or larger telescope will begin to resolve the haze of M3 into hundreds of tiny stellar points surrounding the cluster's moderately compressed nucleus.

By doubling the telescope's aperture to 8 inches (20 cm), stars are resolved across the cluster's face, with several seeming to line up in long strings (Figure 6.17). The famous eighteenth-century astronomer William Herschel saw some of these stars, which he called "star chains." In reality, these chains are nothing more than chance alignments of stars within the cluster as seen from our vantage point.

### M53: Globular Cluster in Coma Berenices

DISTANCE FROM EARTH: 62,000 light-years

FINDING FACTOR: ****

"WOW!" FACTOR:    Binoculars: *    Small telescopes (3″ to 5″): **
Medium telescopes (6″ to 8″): **

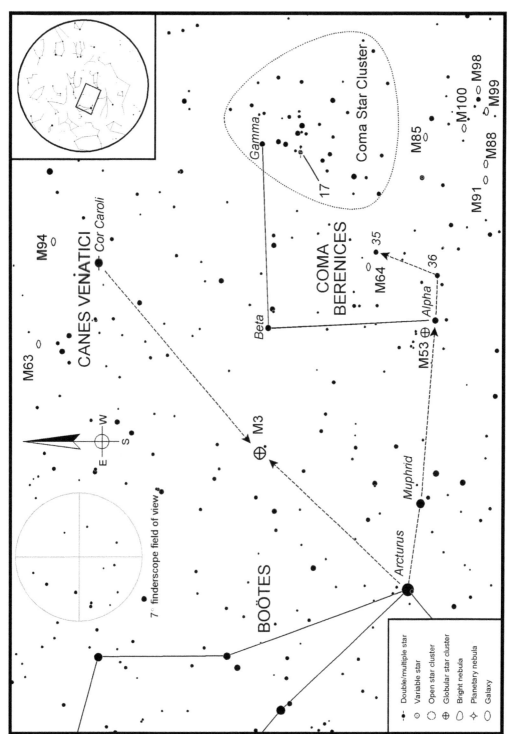

**Figure 6.16** *Spring Sky Window 7 includes the bright star Arcturus in Boötes.*

139

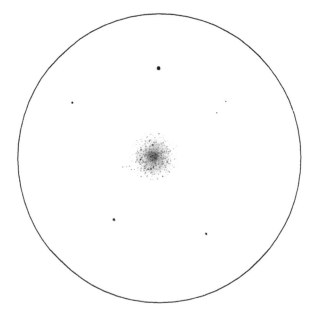

**Figure 6.17** *M3, a grand globular cluster in Canes Venatici, as drawn through the author's 8-inch reflector at 120×. South is up.*

### A Debt to M3

Discovered in 1764 by Charles Messier, M3 is the brightest globular cluster in the northern spring sky and is one of the brightest in the entire sky. More than half a million stars are believed to be held in its grasp, making M3 one of the largest members of the Milky Way's family of globular clusters.

M3 is also the richest globular cluster in terms of variable stars, notably RR Lyrae variables. Like Cepheid variables, RR Lyrae variable stars can be used to determine distances to far-off objects like M3. By studying RR Lyrae variables, astronomers have learned that these stars have a very precise period-luminosity relationship. The more rapidly the star's brightness varies, the greater its inherent luminosity. If you know a star's luminosity as well as its magnitude, it is a simple matter to calculate the star's distance from Earth.

Amateur astronomers everywhere should pay homage to M3, for it was the fuel that lit the urge in Charles Messier to compile his now-famous catalog of deep-sky objects. While Messier had made extensive notes on the objects that later became the first two listings in his yet-to-be-compiled catalog, legend has it that it was the discovery of M3 that led him to his systematic quest for other cometlike imposters.

### Where to Look

From brilliant Arcturus in Boötes, head west to the star Muphrid, often shown as forming the "tail" of the Boötes "kite." Continue westward to 4th-magnitude Alpha Comae, the brightest star in the faint constellation Coma Berenices. M53 is just to Alpha's northeast, in the same low-power eyepiece field.

### What You'll See

**Through Binoculars.** Most binoculars display M53 as a round, nebulous disk drawing to a brighter center. Look for a small interstellar ball of cotton set in an attractive star field.

**Through a Telescope.** Even telescopes in the 3- to 8-inch (7.5- to 20-cm) range have difficulty showing M53 as anything more than a perfectly round fuzz-ball. A few feeble points of light toward the edges and a little graininess to the core can be spotted through 6- and 8-inch (15- and 20-cm) instruments, but even these are tentative.

## M64: Spiral Galaxy in Coma Berenices

NICKNAME: Black-Eye Galaxy
DISTANCE FROM EARTH: 19 million light-years
FINDING FACTOR: ***
"WOW!" FACTOR:    Binoculars: *        Small telescopes (3″ to 5″): **
                 Medium telescopes (6″ to 8″): ***

### Where to Look

To find M64, follow the directions given above for M53. Continue farther west to the star 36 Comae, then hook northwestward to 35 Comae. M64 is just to 35 Comae's northeast, visible in the same low-power eyepiece field.

### What You'll See

**Through Binoculars.** Shining at 8th magnitude, M64 may just be glimpsed with 50-mm binoculars as an oval patch of light with a brighter center. It's a tough catch, so you'll need patience and perseverance to find it.

**Through a Telescope.** Smaller backyard telescopes present M64 (Figure 6.18) as an oval glow with a noticeably off-center stellar nucleus. But wait, there's more. A careful look through a 4-inch (10-cm) or larger telescope at about 100× will begin to show a band of obscuring dust wrapping around the galaxy's core, giving rise to its nickname, the Black-Eye Galaxy. Try averted vision if you have trouble seeing the dark lane directly.

#### Black-Eye Galaxy

M64 was discovered by Edward Pigott in 1779, although the "black eye" wasn't noticed until John Herschel, the son of William, spotted it in 1833 during his survey of his father's catalog. The dark lane is made up of a series of irregularly shaped clouds of dust silhouetted in front of one of the galaxy's spiral arms.

Interestingly, the black eye seems to mark a boundary separating two sections of the galaxy. The inner core, some 6,000 light-years across, is rotating in the opposite direction of the outer region, which extends for more than 40,000 light-years. This bizarre situation, which most likely caused the churned-up dust that forms the black eye, is probably the aftermath of a collision between a small galaxy and a large galaxy that is still in the process of settling down.

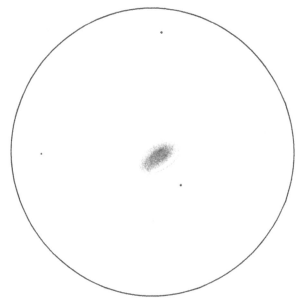

**Figure 6.18** *M64, the Black-Eye Galaxy in Coma Berenices, as seen through the author's 8-inch reflector at 120×. South is up.*

## Coma Berenices Star Cluster: Open Cluster in Coma Berenices

DISTANCE FROM EARTH: 288 light-years
FINDING FACTOR: *
"WOW!" FACTOR:    Binoculars: ****    Small telescopes (3″ to 5″): **
Medium telescopes (6″ to 8″): *

### Where to Look

The Coma Berenices Star Cluster, not to be confused with the Coma-Virgo Galaxy Cluster discussed earlier, takes up the entire western third of that faint constellation. Scan to the west of Arcturus through your binoculars or finder-scope until you come to a wide swarm of stars. You can't miss it.

### What You'll See

**Through Binoculars.** The Coma Berenices Star Cluster is, without a doubt, one of the sky's finest open star clusters for binoculars. Eighty stars of 5th and 6th magnitude spread across 5° (that's right, degrees!) belong to the group. With 7× wide-field glasses, the view is full of fairly bright stars set in gentle arcs and jagged lines in a pattern resembling a V-shaped flock of northward-flying geese.

Several double stars highlight the Coma Berenices Cluster. Of these, 17 Comae is one of the easiest to spot. Even through the smallest opera glasses, an observer will have little trouble distinguishing the 7th-magnitude companion from the 5th-magnitude primary thanks to their wide 2.5 arc-minute separation. Look for the pair just east of the cluster's center.

**Through a Telescope.** Leave the telescopes at home for this cluster. Its wide span compared to the narrow field of most telescopes proves that in the heavens, as here on Earth, sometimes you really can't see the forest for the trees.

# 7

# Summer Sky Windows

Our summer sky (Figure 7.1) is dominated by three brilliant stars—Vega, Deneb, and Altair—set in a large triangle. Known as the Summer Triangle, this is not an official constellation but rather an asterism. Each of the three stars in the Summer Triangle belongs to a separate constellation: Vega in Lyra, Deneb in Cygnus, and Altair in Aquila.

Brightest of the Triangle's stars is Vega, found in the middle of the summer star map and high overhead during the months of July, August, and September. Vega, a brilliant blue-white star glistening like a diamond, belongs to the constellation Lyra, the Lyre. Vega represents part of the Lyre's handle, while four faint stars depict the instrument's main body where the strings are strung. This is the mythical musical instrument of Orpheus, son of the Sun god Apollo. According to legend, Apollo taught Orpheus to play the instrument so beautifully that even savage beasts were soothed into submission. After Orpheus died, his lyre continued to play lyrical music and so was placed in the sky by the gods for all to see.

Leave the Summer Triangle temporarily by drawing an imaginary line between Vega and the springtime star Arcturus, still seen in the northwest. About a third of the way along that line are the faint stars of Hercules the Giant. All are visible only in darker suburban or rural skies. Four stars nicknamed the Keystone create the Giant's torso. Two curved lines of stars extend southward to form his arms, while two other star-curves stretch out to the Keystone's north to form his legs. That's right, the celestial Hercules appears to be standing upside down, apparently with his feet on the head of Draco the Dragon.

# Summer sky

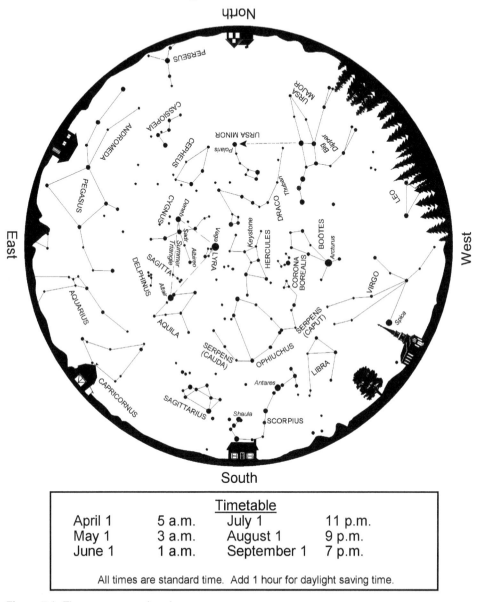

**Figure 7.1** *The summer evening sky.*

| Timetable | | | |
|---|---|---|---|
| April 1 | 5 a.m. | July 1 | 11 p.m. |
| May 1 | 3 a.m. | August 1 | 9 p.m. |
| June 1 | 1 a.m. | September 1 | 7 p.m. |

All times are standard time. Add 1 hour for daylight saving time.

Draco winds its way across the northern summer sky, spanning the gap between Cepheus to the pointer stars of the Big Dipper. Draco is one of the oldest constellations, perhaps dated as far back as ancient Sumeria, and one of the few that actually looks like what it represents. Four faint stars create the Dragon's head, while its long, thin body extends toward Cepheus, coils around Ursa Minor, and slithers between the Bears. Though visible all year long, Draco is most easily seen in the summer. About two-thirds of the way along the body of Draco is the star Thuban. More than 4,500 years ago, when the ancient Egyptians built the pyramids, the Earth's north polar axis was not aimed toward Polaris, as it is today, but rather at this star. This is due to a slow wobbling of the Earth's axis called precession. In 14,000 years, the pole will be aimed toward Vega in the summer constellation Lyra. Twenty-six thousand years from now, Earth's north pole will again be aimed toward Polaris.

Return to the Summer Triangle and the star Altair, its southernmost star. Altair marks the beak of Aquila, the Eagle, while fainter stars denote the out-stretched wings and tail feathers, although it might be easier to see a pterodactyl here than to imagine an eagle.

The third star in the Triangle is Deneb, which marks the tail of Cygnus the Swan. In this case, the shape of a swan captured in midflight is quite easy to imagine. The Swan's body and long neck extend to the star Albireo, while wings stretch outward from either side of the star Sadr in the Swan's body. Many people also refer to this pattern as the Northern Cross for its likeness to a crucifix.

If you are viewing from darker suburban or rural skies, look for a hazy band of light passing through the Summer Triangle and stretching from the northeast to the south. This is our galaxy, the Milky Way—a spiral galaxy made up of about 200 billion stars. Actually all of the stars seen in the sky belong to the Milky Way. But when we look along the edge of our galaxy, we are seeing millions of other, more distant stars that are too faint to make out as individual points of light with our eyes alone. Instead, their light blends to form the lane that we call the Milky Way.

Follow the Milky Way down toward the southern horizon, to the bright star Antares in Scorpius, the Scorpion. Antares is known to be a red super-giant star, nearly 2,000 times larger than our Sun, although only about half as hot. The curve of the Scorpion's body extends toward the southern horizon and ends at the star Shaula, which marks the Scorpion's stinger. According to legend, Scorpius killed the mighty hunter Orion (seen in the winter sky) by stinging him on the foot. To punish the Scorpion for this dastardly deed, the gods banished Scorpius to the sky forever, placing it directly opposite Orion so that it could never harm him again.

Just east of Scorpius is Sagittarius the Archer, who also happens to be a centaur, which is a mythical creature with the upper body of a man and the lower body of a horse. Most artists draw Sagittarius facing toward the right (west), his bow and arrow aimed toward the heart of Scorpius the Scorpion. Perhaps Sagittarius misfired his first arrow, for we see the tiny constellation Sagitta the Arrow lying just above (north of) Altair. Sagittarius may have represented a centaur to our ancestors, but to skywatchers today, its outline is more reminiscent of a stove-top kettle or teapot.

Both Sagittarius and Scorpius are found along the ecliptic, the path in the sky followed by the Sun, the Moon, and the planets. In fact, this is probably how Antares got its name. *Ant-ares* literally translates from Greek as "rival of Ares," a rivalry probably caused by the star's amazing resemblance to Mars when seen with the naked eye. (Ares was the Greek god of war, while Mars was Rome's name for the same deity.)

Though there are twelve constellations assigned to the zodiac, the Sun will actually travel through thirteen each year. That forgotten constellation of the zodiac is Ophiuchus the Serpent Bearer. Ophiuchus fills most of the large void between Antares and Vega. According to legend, Ophiuchus was a great physician. The story goes that Ophiuchus (also known as Aesculapius, the son of Apollo) became such an amazing healer that Pluto, god of the Underworld, complained about his lack of new clientele! Finally, to appease Pluto, Jupiter killed Ophiuchus, but as a lasting memorial Jupiter placed him among the stars. Ophiuchus is shown holding a serpent named Serpens. Serpents and snakes have long been associated with healing, for their periodic molting was looked upon by our ancestors as a rebirth. (This association of snakes with healing is still evident today on the caduceus, which is used as a symbol by the American Medical Association.) Serpens is divided into two parts. Its head, shown on the map as Serpens Caput, extends to the right (west) of Ophiuchus, while its tail, Serpens Cauda, lies to the left, or east.

Continuing our tour of the ecliptic, the next constellation east of Sagittarius is Capricornus the Sea-Goat. The faint stars of the Sea-Goat form a large triangle seen in summer's southeastern sky. A sea-goat is an imaginary creature with the head and front legs of a goat and the tail of a fish. Where this constellation came from is a bit of a mystery, since this bizarre animal is not mentioned anywhere in classical mythology. Some suggest that Capricornus actually represents the mythical prankster Pan. Pan had the head and legs of a goat, but the body of a man.

Just above (north of) the Sea-Goat and to the left (east) of Altair lies the tiny but distinctive constellation of Delphinus the Dolphin. Delphinus is marked by a diamond of four stars, which form the Dolphin's body, and a fifth star below (south) marking its tail. None of the individual stars in Delphinus is very bright, but the constellation's kitelike pattern stands out quite well against its barren surroundings.

Figure 7.2 holds the key to the summer sky windows. The map, a duplicate of Figure 7.1, shows seven highlighted areas, which are discussed in detail below. Each of those windows shows an enlarged area of the sky plotting fainter stars and selected deep-sky objects. Compare the two maps, decide which area you wish to explore tonight, and then head outside to find that region. Once the naked-eye stars and constellations have been located, switch to the close-up sky window found later in this chapter and, using your finderscope or binoculars, begin your journey into the depths of the universe.

Each sky window includes a miniature all-sky chart to show its position relative to the rest of the night sky. To help readers better appreciate the scale

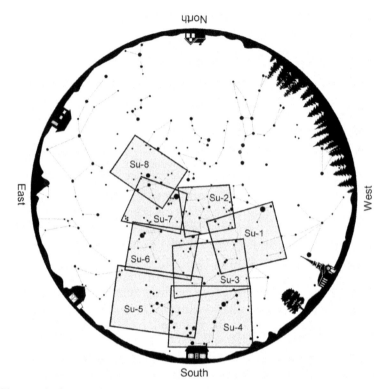

**Figure 7.2** *Summer key map, showing each of the summer sky windows to come.*

of things, a crosshair frame is also included in each window to show the typical field of a 6 × 30 finderscope (about 7°).

## Summer Sky Window 1

The first Summer Sky Window (Figure 7.3) features a single deep-sky object, M5, one of the sky's most spectacular globular clusters. Tucked in an out-of-the-way corner of the early summer sky, M5 is often bypassed by observers in favor of other globulars that are easier to find. But those who take the trouble to hunt it down are rewarded with a striking sight.

### M5: Globular Cluster in Serpens

DISTANCE FROM EARTH: 23,000 light-years
FINDING FACTOR: **
"WOW!" FACTOR:    Binoculars: ***      Small telescopes (3″ to 5″):  ***
                 Medium telescopes (6″ to 8″):  ****

### *Where to Look*

M5 is bright enough to be visible with very little optical assistance, but getting to it may prove challenging at first, since the stars of Serpens are on the

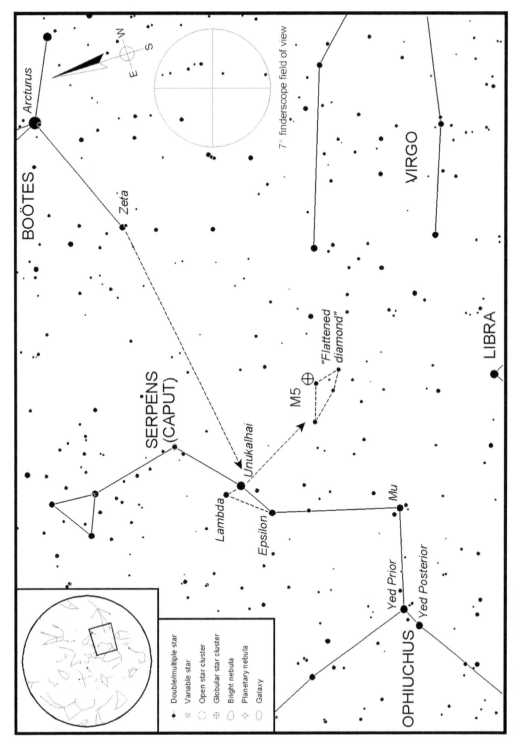

**Figure 7.3** *Summer Sky Window 1, centered on the constellation Serpens Caput, the Serpent's head.*

faint side. Aim your finderscope or binoculars at brilliant Arcturus, the zero-magnitude star in the spring constellation Boötes, which should still be visible in the western sky. Slide to the southeast, to the fairly faint (4th magnitude) star Zeta Boötis, then continue twice the distance along a slightly more easterly course until you come to a not-quite-right triangle of stars. The triangle's brightest star, positioned at the not-quite-right angle, is named Unukalhai and shines at magnitude 2.6. Turn to the southwest and look for a faint pattern of four dim stars in the shape of a flattened diamond. M5 lies just north of the diamond.

### What You'll See

**Through Binoculars.** Once you are in the right neighborhood, M5 is easy to spot as a nebulous globe through just about all binoculars. Like most globular clusters, M5 will look perfectly circular, highlighted with a brighter core.

**Through a Telescope.** Through small telescopes, M5 will look like a ball of cotton. But on closer inspection with a 4-inch (10-cm) telescope (maybe even a 3-inch [7.5-cm] with great optics), you can begin to resolve some of the cluster's individual stars. An 8-inch (20-cm) telescope reveals a multitude of stellar points across the entire disk. Many of the stars in the cluster seem to form lines or strings.

---

**M5**

M5 was discovered by Gottfried Kirch in May 1702, although little attention was paid to it until Messier independently found it in 1764. Modern studies show that M5 is one of the Milky Way's oldest globular clusters. Astronomers can tell this by analyzing light from its individual stars. They have found that all of the cluster's stars have begun to evolve into red giants, leading many to estimate the cluster's collective age as perhaps 13 billion years.

Estimates say that M5 may contain as many as one million stars, all of which are crammed into a space about 130 light-years across. This makes M5 one of the largest globulars known.

---

## Summer Sky Window 2

Just about any astronomy book that discusses globular clusters is bound to illustrate them with the granddaddy of all globulars, M13, the Great Hercules Cluster. That's the focus of Summer Sky Window 2 (Figure 7.4) but is by no means the only object of interest in the area. The dim stars of Hercules, framed between the brilliant stars Arcturus and Vega, also hold a second globular, a planetary nebula, and one of the season's prettiest binary stars.

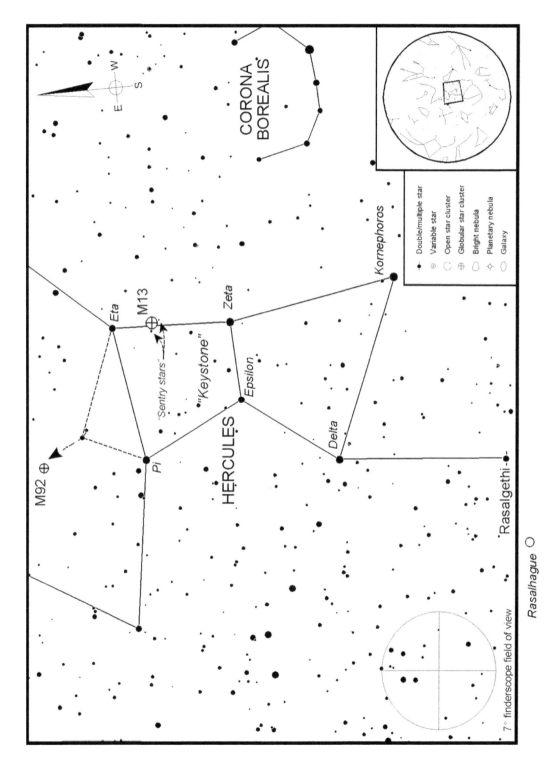

**Figure 7.4** Summer Sky Window 2 features one of the season's showpieces, M13, as well as several lesser-known targets.

## M13: Globular Cluster in Hercules

NICKNAME: Great Hercules Cluster
DISTANCE FROM EARTH: 25,000 light-years
FINDING FACTOR: *
"WOW!" FACTOR:    Binoculars: ***    Small telescopes (3″ to 5″): ****
                 Medium telescopes (6″ to 8″): ****

### Where to Look

To find M13, first find its home constellation, Hercules. Although the stars of this celestial giant are rather faint, their location in the sky can be found by drawing an imaginary line between the bright stars Vega in Lyra, to their east, and Arcturus in Boötes, to their west. Hercules dwells about one-third of the way from Vega to Arcturus.

M13 lies along the western side of the Hercules Keystone, about a quarter to a third of the way between the stars Eta and Zeta Herculis. Train your finderscope or binoculars in that general direction, and keep an eye out for two faint "sentry stars" that stand at attention to either side of M13.

On exceptionally dark, transparent nights, it is possible to see M13 with the eye alone, especially if it is high in the sky, away from the haze of the horizon.

### What You'll See

**Through Binoculars.** Binoculars readily show M13 as a fluffy piece of celestial cotton suspended against the firmament. With a keen eye, you will be able to see that it is clearly not a star but rather a fuzzy patch of light highlighted by a brighter core.

**Through a Telescope.** M13 is a beautiful sight through any telescope (Figure 7.5) and is one of the sky's true showpiece objects. The smallest telescopes show a fuzz ball around a bright central core. A 3.5- or 4-inch (9- or 10-cm) telescope will begin to show that there is more—much more—to M13 than just a nebulous glow, as the cluster's edges begin to dissolve into myriad distant suns. Try about 100× for a good view. Double the aperture and M13 erupts into a huge globe of tiny points, almost as if someone dropped a pinch of sugar onto a black-velvet backdrop. If you have an 8-inch (20-cm) or larger telescope, look carefully and see if you notice how the outer cluster members form chains, or lines, radiating outward from the core.

### M13: One of the Finest Globular Clusters

Edmond Halley, famous for the comet that bears his name, was the first person to lay eyes on M13 when he spotted it in 1714. Messier subsequently added it to his list several decades later, but he only described it as a starless nebula, since his telescopes were incapable of showing any of its stars. That task was left to William Herschel. In fact, Herschel also coined the phrase "globular cluster" in 1789. M13, the finest example of a globular cluster found anywhere north of the celestial equator, contains up to one million stars within its 165-light-year diameter.

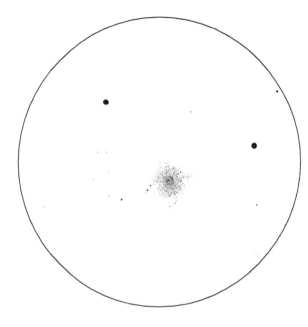

**Figure 7.5** *M13, as drawn through the author's 8-inch reflector at 55×. Note the faint galaxy NGC 6207 off to the lower right of the field. South is up.*

## M92: Globular Cluster in Hercules

DISTANCE FROM EARTH: 26,400 light-years
FINDING FACTOR: **
"WOW!" FACTOR:    Binoculars: **      Small telescopes (3″ to 5″): ****
                 Medium telescopes (6″ to 8″): ****

### Where to Look

Once you find Hercules, M92 can be located by connecting with an imaginary line the stars Pi and Eta Herculis, the two dim stars that mark the northern side of the Hercules Keystone. Both stars may just squeeze into your finder-scope's field of view. Take a look for a slightly fainter star to their north. These three stars should form a fairly conspicuous right triangle through your finder. From that third star, which marks the triangle's right angle, head north-northeast about a third of a finder field to M92. Although dimmer than M13, M92 should be bright enough to show as a dim "star" in 6 × 30 and larger finderscopes, unless light pollution is severe.

### What You'll See

**Through Binoculars.** Binoculars show M92 as a misty glow that draws to a brighter central core. Wide-angle glasses can just squeeze both M92 and M13 into the same 10° field for a direct comparison between the two.

**Through a Telescope.** M92 is a beautiful globular cluster—a true showpiece—that often goes unappreciated because of M13's even more spectacular nature.

One look will convince you that finding M92 is certainly worth the effort! A 4-inch (10-cm) telescope is enough aperture to begin resolving some of the stars around the edges of this grand globular, while doubling the aperture more than doubles the number of stars resolved. In either case, since M92 is about half the size of M13, be prepared to crank up the magnification to about 100× to 150× for the best view.

### Rasalgethi: Binary Star in Hercules

DISTANCE FROM EARTH: 382 light-years
FINDING FACTOR:  *
"WOW!" FACTOR:    Binoculars: *       Small telescopes (3″ to 5″): ****
                 Medium telescopes (6″ to 8″): ****

#### Where to Look

Even though Rasalgethi, also known as Alpha Herculis, is one of the brightest stars in Hercules, it only shines at 3rd magnitude, making it difficult to pinpoint from light-polluted skies. For me, the easiest way to locate it is to first identify the outline of the neighboring constellation Ophiuchus. Zero in on the star Rasalhague at the top star in the Ophiuchus pentagon, then look just to its west for dimmer Rasalgethi.

#### What You'll See

**Through Binoculars.** While binoculars can't resolve the two stars that make up this distant binary system, they might show a slightly warm, golden tint to Rasalgethi.

**Through a Telescope.** Rasalgethi is a true gem of the summer sky. Even a 2.4-inch (6-cm) refractor will reveal the brighter "A" star, an orangish 3.5-magnitude sun, teamed with a white 5.4-magnitude companion some 5 arc-seconds to the southeast. The companion may look slightly bluish, even though it is not a blue star. The false coloring is probably caused by the contrast between it and the primary sun.

## Summer Sky Window 3

You've entered globular country! As Summer Sky Window 3 (Figure 7.6) shows, no constellation holds more Messier globulars than the large, faint outline of Ophiuchus. Most of its globulars are quite bright and so are reasonably easy to find, although M107 may give you a run for your money.

### M10: Globular Cluster in Ophiuchus

DISTANCE FROM EARTH: 13,400 light-years
FINDING FACTOR:  **
"WOW!" FACTOR:    Binoculars: **      Small telescopes (3″ to 5″): ***
                 Medium telescopes (6″ to 8″): ***

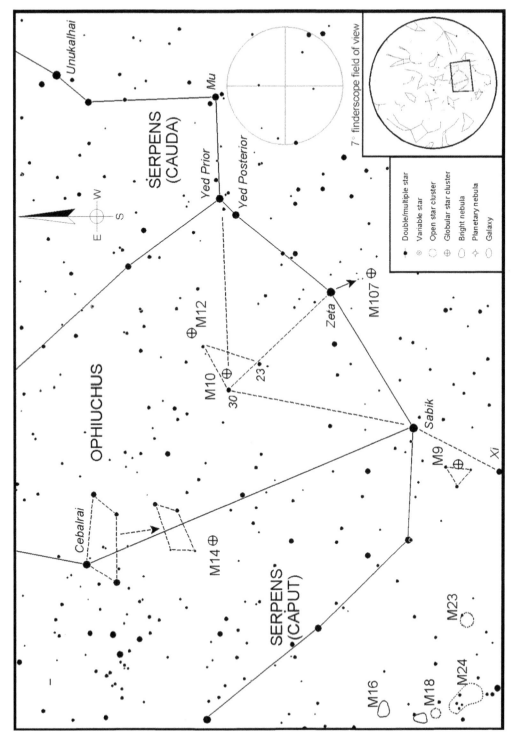

**Figure 7.6** *Summer Sky Window 3 holds the key to five of the Messier catalog's twenty-nine globular clusters.*

### Where to Look

M10 and M12, detailed below, lie within the main body of Ophiuchus. Unfortunately, Ophiuchus itself can be difficult to find at first, let alone these dim objects within! Let's begin with the big picture, then work our way in. Ophiuchus is most easily located by stopping halfway between the bright stars Antares in Scorpius and Vega in Lyra. From dark, rural skies, you should see the grayish lane of the Milky Way just skirting the constellation to the east.

To find M10, first find the stars that outline the southern boundary of Ophiuchus, just north of the head of Scorpius. The western end is marked by two stars—Yed Prior and Yed Posterior—while the eastern end is bounded by Sabik. The line kinks a little to the south, to the star Zeta Ophiuchi. Looking at Figure 7.6, notice the pair of identical triangles drawn using these stars along with the faint sun 30 Ophiuchi. Though 30 may just be visible to the naked eye if your skies are very dark, more than likely, you'll need to find it with your finderscope. Look at the chart, then look at the sky and aim your finder toward where 30 should be. You might need to scan a little but hopefully not too much before you bump into it. M10 lies just to the west of 30 and should be visible through the finder.

### What You'll See

**Through Binoculars.** With most binoculars, M10 looks like a relatively dim patch of grayish light, set in a nice star field. To the west, M12 should also be visible in the same field of view.

**Through a Telescope.** M10 is a fairly apparent target through amateur telescopes. Some of its multitude of stars are resolvable through 6-inch (15-cm) telescopes, but you will need at least an 8-inch (20-cm) for a good view. The cluster's bright, central core looks somewhat oblong, or shaped like a kidney bean, toward the southwest.

## M12: Globular Cluster in Ophiuchus

DISTANCE FROM EARTH: 17,600 light-years
FINDING FACTOR: **
"WOW!" FACTOR:     Binoculars: **     Small telescopes (3″ to 5″): ***
                  Medium telescopes (6″ to 8″): ***

### Where to Look

Pull back from M10 and look through your finderscope. Can you see how M10 lies within a triangle of stars, made up of the stars 23 and 30 Ophiuchi as well as an anonymous star to their west? All fit pretty comfortably into the same finder field. Take aim at that western star, then switch back to your telescope. M12 should be in the western half of the field and is also probably visible directly through the finder.

**Figure 7.7** *M12, one of Ophiuchus's finest targets. South is up in this photograph by George Viscome through a 14.5-inch f/6 reflector.*

### What You'll See

**Through Binoculars.** M12 looks nearly identical to M10 through binoculars. Both fit neatly into the same binocular field, so direct comparison is easy. Study them both for similarities and differences. You might find that the core of M12 is a little less evident than that of M10, but otherwise they are twins.

**Through a Telescope.** M12 (Figure 7.7) appears a bit smaller and has a slightly looser stellar concentration than M10, making resolution possible, at least in part, through a 4-inch (10-cm) telescope. Try a fairly high power, between 120× and 150×, for the best view.

## M107: Globular Cluster in Ophiuchus

DISTANCE FROM EARTH: 19,600 light-years
FINDING FACTOR: ****
"WOW!" FACTOR:    Binoculars: *      Small telescopes (3″ to 5″): **
                 Medium telescopes (6″ to 8″): **

### Where to Look

Once you locate the imposing, though dim body of Ophiuchus, center your attention (and your finderscope or binoculars) on Zeta Ophiuchi. In the same finder or binocular field, look for a tiny right triangle of faint suns just to the star's south. The southward pointing side of the triangle is aimed directly at M107, which might also be visible in 50-mm and larger finderscopes and binoculars.

### What You'll See

**Through Binoculars.** Through 7× and 10× binoculars, M107 can be spotted with some effort as a faint, diffuse glow with a brighter core, just to the south of that small right triangle of stars. Although it is one of the sky's more loosely packed globular clusters, there is no hope of resolving any of its dim suns through binoculars.

**Through a Telescope.** Resolving this globular's stars through most backyard telescopes is nearly impossible as well. Instead, M107 looks like little more than a slightly oval blur. Its stars are so dim that the light-gathering ability of at least a 10-inch (25-cm) telescope is needed to make out a few weak points of light.

## M14: Globular Cluster in Ophiuchus

DISTANCE FROM EARTH: 27,000 light-years
FINDING FACTOR: ***
"WOW!" FACTOR:    Binoculars: **    Small telescopes (3″ to 5″): ****
                 Medium telescopes (6″ to 8″): ****

### Where to Look

Although M14 is relatively bright at 7th magnitude, it is located far from any handy star-hopping candidates, so it may take a little longer to bump into it than you might like. Let's begin at the star Cebalrai, marking the upper left (northeast) shoulder of Ophiuchus. Find the three other nearby stars that combine with Cebalrai to make the parallelogram shown on Summer Sky Window 3. Now, drop to the south about one finder field to a second, fainter parallelogram. M14 is a little farther to the south, near a close-set pair of 7th- and 9th-magnitude field stars (neither of which is shown on the chart, as they are below its magnitude limit).

### What You'll See

**Through Binoculars.** M14 is a relatively bright object that stands out well in a rich field of stars. As through Messier's telescope, binoculars reveal it as a vague smudge without distinct stars.

**Through a Telescope.** M14 appears smaller than M10 and M12 but larger and brighter than M107. Most backyard telescopes won't show it to be anything more than a ball of celestial fluff, since none of its more than 100,000 stars are brighter than about 15th magnitude, which is fainter than Pluto. Therefore, only telescopes larger than 10 inches (25 cm) stand much chance of seeing anything beyond a round disk that grows brighter toward its center.

## M9: Globular Cluster in Ophiuchus

DISTANCE FROM EARTH: 26,000 light-years
FINDING FACTOR: **
"WOW!" FACTOR:    Binoculars: *    Small telescopes (3″ to 5″): **
                 Medium telescopes (6″ to 8″): **

### Where to Look

Locate the star Sabik, which marks the southeastern corner in Ophiuchus's large, hexagonal body. Look through your finderscope for the star Xi Ophiuchi, which should also be visible faintly to the unaided eye from darker suburban and rural yards. Just east of the halfway point between Sabik and Xi, find a right triangle of dim stars. M9 lies along the triangle's hypotenuse.

### What You'll See

**Through Binoculars.** M9 appears quite small in binoculars, certainly smaller than neighboring M10 and M12. Its intense core appears surrounded by a fuzzy glow.

**Through a Telescope.** The small apparent size of M9 means that higher magnifications will give the best view. Once you center it in the field of view, switch to between 100× and 150×. Three- to 6-inch (7.5- to 15-cm) telescopes should display a slightly oval central core surrounded by a circular outer halo. Eight-inch (20-cm) and larger telescopes can partially resolve the cluster by adding a few feeble points of light around the edges of M9.

> ### M9
>
> M9 was discovered by Charles Messier on May 28, 1764. The slightly oval appearance mentioned above is likely caused by interfering clouds of dust, which are especially dense toward the cluster's north and west. Were it not for these dust clouds, M9 would probably look round like most other globulars and would be at least one magnitude brighter. M9 measures 70 light-years across.

## Summer Sky Window 4

Scraping the southern horizon from midnorthern latitudes, Summer Sky Window 4 (Figure 7.8) opens on the constellation Scorpius, a veritable playground for stargazers. There's the bright red star Antares, the gentle glow of the Milky Way, and some wonderful star fields that were made for a leisurely scan through binoculars on warm summer nights. And in among those stars, we find many wonderful deep-sky objects.

## M4: Globular Cluster in Scorpius

DISTANCE FROM EARTH: 7,200 light-years
FINDING FACTOR: *
"WOW!" FACTOR:    Binoculars: **    Small telescopes (3″ to 5″): ****
                 Medium telescopes (6″ to 8″): ****

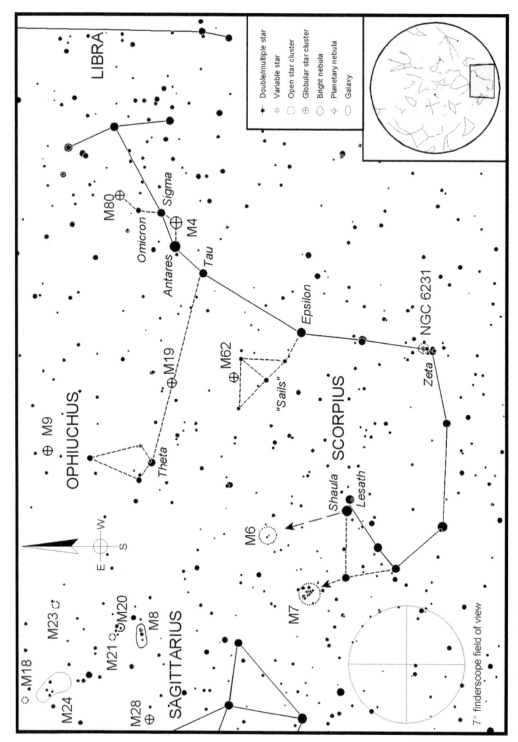

**Figure 7.8** *Summer Sky Window 4 frames the southern summer constellation Scorpius the Scorpion.*

## Where to Look

M4 is easy to spot, since it lies just over a degree to the west of the red supergiant star Antares. You'll find it forming a triangle with Antares to its east and 3rd-magnitude Sigma Scorpii to its northwest. All three readily fit into the same finderscope field, so M4 should be easy to pick out. The cluster's low altitude from midnorthern latitudes, however, makes a good view to the south an absolute must.

## What You'll See

**Through Binoculars.** Even small binoculars will be able to show M4 as a tiny, undefined smudge of grayish light just to the west of Antares.

**Through a Telescope:** M4 is one of the most interesting and unusual globular clusters visible through backyard telescopes. At first pass it may just look like a round blur, but

### M4

One of the closest globular clusters to Earth, M4 was discovered by Philippe Loys de Cheseaux in 1746. In all, more than 100,000 stars crammed within about a 100-light-year diameter call M4 home.

take a closer look. You should see that M4 is more concentrated in the center and that its core is bisected by a brighter line of light. This odd appearance, evident in Figure 7.9, should be visible in telescopes as small as 3 inches (7.5 cm) aperture. Try a moderate magnification, say around 75× to 100×, for the best view. Larger telescopes show that this unusual feature is actually a chain of 11th-magnitude cluster stars that just happen to line up in a row.

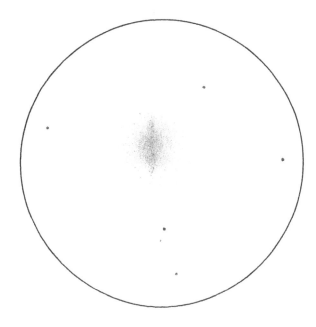

**Figure 7.9** *Globular cluster M4 as drawn through the author's 8-inch reflector at 55×. Note the prominent "star bar" running through the cluster's center. South is up.*

## M80: Globular Cluster in Scorpius

DISTANCE FROM EARTH: 27,400 light-years
FINDING FACTOR: **
"WOW!" FACTOR:    Binoculars: **      Small telescopes (3″ to 5″): ***
                 Medium telescopes (6″ to 8″): ****

### Where to Look

Once you find M4, move a little farther to the north, to the star Sigma Scorpii. Snake along a winding road of stars to the north, to 5th-magnitude Omicron Scorpii, then onward to M80, tucked just to Omicron's northwest.

### What You'll See

**Through Binoculars.** Given even reasonably dark skies, binoculars easily detect the dim (7th-magnitude) disk of M80. At first it might look like just another point of light set in a dense star field, but if you concentrate on it, you should be able to make out its round, softly glowing disk highlighted by a more intense center.

**Through a Telescope.** Although that round blur grows in brightness and size, even telescopes as large as 8 inches (20 cm) aperture have a tough time resolving M80's individual stars, since none shine brighter than 14th magnitude. Instead you will probably notice a "roughness" or "mottling" in its central core. Try using a relatively high magnification (say, around 150×), but you will have to wait for steady sky conditions, since it is so low in our southern sky.

## M19: Globular Cluster in Ophiuchus

DISTANCE FROM EARTH: 27,000 light-years
FINDING FACTOR: **
"WOW!" FACTOR:    Binoculars: *     Small telescopes (3″ to 5″): **
                 Medium telescopes (6″ to 8″): **

### Where to Look

M19 lies to the east of the bright star Antares and the curve of the Scorpion's body. To find it, first spot 3rd-magnitude Tau Scorpii just to the southeast of Antares. Trace a line to the northeast, to the star Theta Ophiuchi, the brightest star in an asterism of four stars that looks like a kite. About two finder fields separate Tau and Theta, with M19 lying halfway in between. If you put Theta on the eastern edge of your finder field, then M19 will be just inside the western edge. Or if you prefer, place Tau on the finder's western edge, and M19 will be to the east.

### What You'll See

**Through Binoculars.** M19 is bright enough to be seen through just about any pair of binoculars.  In fact 7× binoculars will not only show M19, but they will also reveal that its fuzzy disk is noticeably oval.

**Through a Telescope.** The unusual shape of M19 is further delineated through telescopes, with the cluster's long dimension (what astronomers call its *major axis*) oriented north-to-south. M19's central core is also unusual in that it is positioned off-center to the north. Unfortunately, the stars in M19 are quite faint, making resolution through 4- and 6-inch (10- to 15-cm) telescopes impossible. A few scattered points of light around the edges might be just visible through 8-inch (20-cm) telescopes, but only on clear, dark nights.

## M62: Globular Cluster in Ophiuchus

DISTANCE FROM EARTH: 21,500 light-years
FINDING FACTOR:  ***
"WOW!" FACTOR:    Binoculars: *      Small telescopes (3″ to 5″): **
                 Medium telescopes (6″ to 8″): **

### Where to Look

M62 is located in between the curve of Scorpius and the teapot asterism in Sagittarius to the east. While no bright stars lie nearby, observers can find their way to this globular cluster by beginning at the 2nd-magnitude star Epsilon Scorpii. Look through your finderscope for two back-to-back right triangles to that star's northeast. Some might imagine them as a pair of sails. M62 lies to the east of the apex and should glow brightly enough to be faintly visible through 30-mm finders.

### What You'll See

**Through Binoculars.** Through binoculars, globular cluster M62 looks like a small, fuzzy blotch of light set in an absolutely stunning star field. Be sure to take your time scanning the region, as there are a number of interesting star clumps and knots in all directions.

**Through a Telescope.** The stars that make up M62 lie beyond the resolution threshold of 4- to 6-inch (10- to 15-cm) telescopes. A few may just be spotted through an 8-inch (20-cm) instrument. Notice how M62 looks oblong, which studies show to be an illusion caused by cosmic dust that partially obscures the cluster.

## NGC 6231: Open Cluster in Scorpius

DISTANCE FROM EARTH: 5,900 light-years
FINDING FACTOR:  *
"WOW!" FACTOR:    Binoculars: ***      Small telescopes (3″ to 5″): ****
                 Medium telescopes (6″ to 8″): ****

### Where to Look

As long as you can see the curve of the Scorpion's body, you can find NGC 6231. Follow its fishhook shape southward until it starts to hook east. Aim at the star marking that eastward turn, 3rd-magnitude Zeta Scorpii. Switch to your telescope and, using your lowest-power eyepiece, look a Full Moon's diameter (about half a degree) to Zeta's north for NGC 6231.

### What You'll See

**Through Binoculars.** NGC 6231 is a tight knot of about 120 stars crushed into a small area of sky spanning only 15 arc-minutes across. Many of its stars shine between 6th and 8th magnitude and are easily seen in all binoculars.

This entire region of Scorpius is just amazing to scan with binoculars! Stars of all colors contrasting against an inky black sky are strewn across a binocular's field of view. Sometimes experienced amateurs become so accustomed to looking for specific objects that they tend to ignore the larger scene. Don't make that mistake here; pause for a moment and absorb the beauty that this corner of the heavens has to offer.

From a dark-sky observing site, the area immediately surrounding Zeta Scorpii and NGC 6231 looks like a comet. Zeta, actually a tightly set trio of stars, marks the false comet's heart, while the dim glow of NGC 6231 and some adjacent stars appear to form a tail fanning out to the north.

**Through a Telescope.** What a striking star cluster! When viewed with a low-power telescope, NGC 6231 displays about a quarter of its 120 stellar baubles shining between magnitudes 5 and 13, with the remaining, fainter suns creating a triangular wedge of celestial fog. Eight-inch (20-cm) and larger apertures blow some of the cloudiness away, revealing even more stars.

### A Collection of Hot, Young Stars

When you gaze at NGC 6231, you are looking at one of the brightest collections of hot, young stars seen anywhere in our galaxy. Sounds a little like a Hollywood promotional ad, but in this case it's true. The cluster is estimated to be only 3.2 million years old, making it a true stellar nursery. Most of its stars are blue-hot spectral type O and type B, each radiating thousands of times the energy of our Sun.

If we could somehow magically reduce the distance to NGC 6231 to only 400 light-years, the same distance as the Pleiades, the brightest half dozen stars in NGC 6231 would outshine all other nighttime stars, even blazing Sirius, in our sky.

## M6: Open Cluster in Scorpius

NICKNAME: Butterfly Cluster
DISTANCE FROM EARTH: 2,000 light-years
FINDING FACTOR: *
"WOW!" FACTOR:    Binoculars: ****    Small telescopes (3″ to 5″): ***
                 Medium telescopes (6″ to 8″): **

### Where to Look

From Antares, follow the curve of the Scorpion all the way around to the south and east until you come to the stars Shaula and Lesath, the two stars that mark the poisonous stinger. Place this pair along the southern edge of your finderscope's view, then look to the north. Can you see a rectangular blur of faint stars? That's M6. In fact, the cluster is so bright that it just might be visible to the naked eye if the sky in that direction is dark and clear.

### What You'll See

**Through Binoculars.** Many people call M6 the Butterfly Cluster because, if you look carefully, you just might see two "wings" of stars spreading out from the cluster's more densely packed center. Seven-power glasses resolve about thirty stars in M6, while giant 11× binoculars add another dozen faint points of light. Although most stars in M6 appear whitish or blue-white, the brightest member star is an orange stellar ember found east of the cluster's center.

**Through a Telescope.** Your binoculars or even finderscope will probably show the cluster better than your telescope. Even at low power, many telescopes have fields of view that are just too confining to take in the full glory of M6.

#### The Butterfly Cluster

Giovanni Hodierna discovered M6 sometime before 1654, although it may have been discovered long before that, as was neighboring M7. M6 is made up of about eighty stars and spans some 20 light-years. Collectively between 50 and 100 million years old, M6 is made up of mostly spectral type O and type A stars. Several of its more massive stars, however, have evolved into yellow, orange, and red giants. In fact, as noted above, the brightest star in M6 is an orange giant star known as BM Scorpii, an irregular variable that fluctuates very slowly and erratically from 7th to 9th magnitude across an average of 850 days.

## M7: Open Cluster in Scorpius

NICKNAME: Ptolemy's Cluster
DISTANCE FROM EARTH: 1,000 light-years
FINDING FACTOR: *
"WOW!" FACTOR:   Binoculars: ****     Small telescopes (3″ to 5″): ***
Medium telescopes (6″ to 8″): **

### Where to Look

M7 lies in the same finder and binocular field as M6. You can't miss it, since it's even bigger and brighter! Start at the star Shaula, which forms a right triangle with two other nearby stars to the east and southeast. Extend the eastern side of the triangle one length to the north, where you will find M7. M7 is also visible to the naked eye on clear, dark nights.

### What You'll See

**Through Binoculars.** When viewed with even a modest pair of opera glasses, M7 bursts into an exceptionally beautiful array of stars spanning more than a Full Moon's diameter. Of the eighty stars identified as cluster members, more than thirty are brighter than 10th magnitude and are visible in binoculars. My 7 × 50 wide-angle binoculars create a three-dimensional effect, as many of the brighter members appear almost floating in front of a field strewn with fainter points of light. There are colors galore in the stars of M7, with several showing subtle hues of yellow and blue. Brightest of all is a yellow star lying close to the group's center.

M6 and M7 are my two favorite binocular clusters of the season and will no doubt become two of yours, as well. When viewed at low power with their surroundings, both combine to create a truly star-spangled sight that is tough to equal anywhere in the sky. One especially memorable view of these clusters came a few summers ago through my 16 × 70 binoculars when the clusters were positioned just above a distant horizon of pine trees. The glittering stars combined with the trees to create a three-dimensional effect that could never be captured in a photograph.

**Through a Telescope.** Like M6, M7 really loses its luster through most telescopes. It's just too big! The full span of M7 is broader than the Full Moon, so unless you are viewing through a short-focus telescope, such as one of the popular 3.1- or 4-inch (8- or 10-cm) "short-tubed" refractors, the cluster will not fit wholly into the field of view. Even then, a low-power eyepiece is a must for the best view. A 4-inch (10-cm) telescope will show about twenty stars spread out across the cluster.

### Ptolemy's Cluster

Records show that M7 was known long before the invention of the telescope. Its discovery traces back to the second-century astronomer Ptolemy, who mentioned it as a nebulous patch in his monumental work *Almagest*. Sixteen centuries later, Charles Messier included it in his now-famous catalog, where it remains as his list's southernmost entry. Messier's description of M7 reads in part "a cluster considerably larger than the preceding (M6)." To me, this seems a bit sterile for such a magnificent family of stars.

In all, eighty stars scattered across 20 light-years form the group. Many shine with a yellowish cast, while others appear blue-white. M7 is about 220 million years old, which is nearly twice as old as many other youthful open clusters including neighboring M6.

## Summer Sky Window 5

Set just to the east of the previous window, Summer Sky Window 5 (Figure 7.10) centers on the constellation Sagittarius. You should be able to recognize its distinctive "teapot" shape from the all-sky map (Figure 7.1).

When we look toward Sagittarius, we gaze at the densely packed center of our Milky Way galaxy, with scores of deep-sky objects. Beautiful globular clusters, sparkling open clusters, and glowing clouds of nebulosity are all found in this wonderful panorama of the universe.

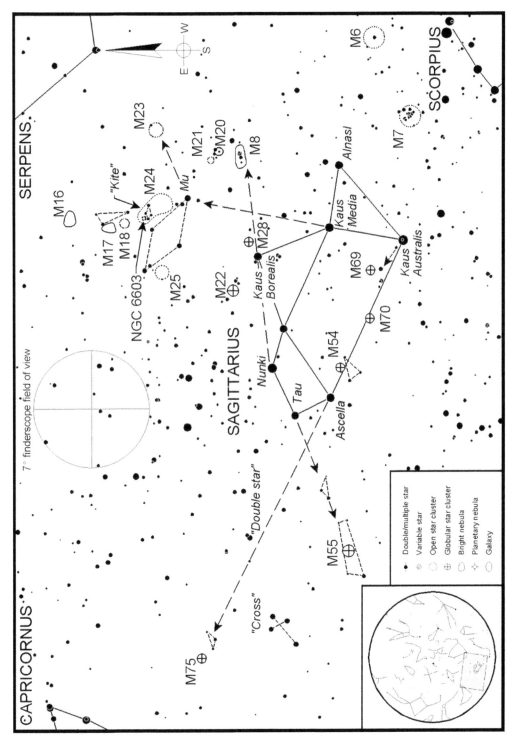

**Figure 7.10** *Summer Sky Window 5 looks toward the constellation Sagittarius and the center of the Milky Way galaxy.*

## M8: Bright Nebula in Sagittarius

NICKNAME: Lagoon Nebula
DISTANCE FROM EARTH: 5,200 light-years
FINDING FACTOR: *
"WOW!" FACTOR:    Binoculars: ****    Small telescopes (3″ to 5″): ****
                 Medium telescopes (6″ to 8″): ****

### Where to Look

To locate M8, the famous Lagoon Nebula, first make a spot of celestial tea by locating the Teapot asterism. Hop from the star Nunki in the Teapot's handle to Kaus Borealis, the star at the tip of the top and hop an equal distance farther to the west-northwest. There you should see M8 through your finderscope, looking like a hazy "star" nestled in the surrounding star clouds. In fact, M8 is visible to the unaided eye as a dim "island" amid the "river" of our galaxy on dark, moonless nights.

What a shame that M8 rides so low in our skies from midnorthern latitudes. From 35° north, for instance, it never rises more than about a third of the way up in the southern sky. That, combined with the haze that often accompanies summer nights, makes M8 a victim of circumstance, since bright nebulae require dark, clear skies to be fully appreciated.

### What You'll See

**Through Binoculars.** M8 is the preeminent bright nebula of the summer sky. There just isn't any competition. Binoculars display a glowing cloud surrounding several stars. Many of those stars belong to the star cluster NGC 6530, which was formed from the nebula. From light-polluted locations, you might only see some of the cluster stars along with the vaguest hint of nebulosity. From dark, rural skies, however, the glowing clouds completely engulf the cluster stars.

**Through a Telescope.** Amateur telescopes display M8 (Figure 7.11) as an iridescent cloud of great intricacy sliced in half by a dark lane, or "lagoon," of obscuring dust that is visible in the smallest amateur telescopes. Viewing M8 through 8-inch (20-cm) and larger instruments, the nebula more than fills the field of a low-power eyepiece and is just breathtaking in its intricacy. You'll see numerous pockets of dark nebulosity twisting in and around patches of bright clouds as well as dozens of cluster stars. Since the Lagoon Nebula spans the width of three Full Moons, a wide field of view is a must to take it all in.

Use that old astronomer's trick, averted vision, to see M8 at its best. A narrowband light-pollution reduction filter (LPR filter, also known as a nebula filter) or an Oxygen-III filter will also help bring out more of M8's subtle nuances.

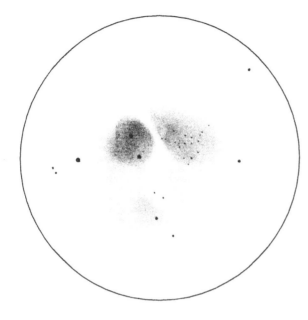

**Figure 7.11** *M8, the Lagoon Nebula, as drawn through the author's 8-inch reflector at 44× and aided by a contrast-enhancing narrowband light-pollution filter. South is up.*

### *Lagoon Nebula*

Estimates say that M8 measures 140-by-60 light-years. The hydrogen gas within M8 is ionized, which means that it is energized into fluorescence by the power of embedded stars. Deep within the nebula, stars are being formed, causing strong stellar winds that create funnel-shaped clouds, which resemble tornadoes here on Earth. Although they remain hidden to amateur telescopes, the Hubble Space Telescope reveals them in amazing detail, each measuring about half a light-year in length.

One of the remarkable features of the Lagoon Nebula is the presence of small, dark pockets known as "evaporating gaseous globules," or EGGs. These clouds, each measuring some 10,000 astronomical units across, are thought to be solar systems in formation. Studies show that there is enough "star stuff" inside the Lagoon Nebula to create as many as 1,000 stars the mass of our Sun.

## M20: Bright Nebula in Sagittarius

NICKNAME: Trifid Nebula
DISTANCE FROM EARTH: Somewhere between 5,200 and 9,000 light-years
              (not well established)
FINDING FACTOR: **
"WOW!" FACTOR:    Binoculars: **      Small telescopes (3″ to 5″): ***
                 Medium telescopes (6″ to 8″): ***

### Where to Look

M20 lies just north of M8. Look for a small diamond pattern of stars that appears a little misty. M20 surrounds the brightest (westernmost) star in the diamond.

### What You'll See

**Through Binoculars.** Most binoculars reveal M20 as a small, relatively faint puff of nebulosity. Although the nebula itself is fairly obvious on transparent nights, the hazy, lazy evenings of summer often makes it tough to spy. Look for a fuzzy spot just to the southeast of that 6th-magnitude star in the diamond.

**Through a Telescope.** At first glance, M20 will probably impress you as a circular patch of grayish light, but there is certainly much more here than just that. Switch to a moderate magnification (say, 75× or so) and take a closer look. The circular patch, vague at first, will take on a whole different personality. Weaving throughout the cloud are distinct dark "lanes" that appear to divide the cloud into thirds. This characteristic led John Herschel to christen M20 the "Trifid," which comes from the Latin word *trifidus*, meaning "split into three." To me, the nebula looks like a giant interstellar pansy.

Far from city lights, even a 2.4-inch (6-cm) telescope will show the three main dark rifts radiating out from the center of M20. Averted vision will certainly help, as will a narrowband light-pollution reduction filter, even if the skies are dark. But be forewarned that since M20 never gets very high above the horizon from midnorthern latitudes, a clear southern horizon free from interfering light pollution is a must for the best view.

You should also notice that there are several stars within M20. Some are simply foreground objects, while others actually lie inside the nebula. Take a look near the center for the multiple star HN 40. The primary star in the system shines at magnitude 7.6, while an 8.7-magnitude companion star lies just to its south. A third, 10th-magnitude companion also shines faintly nearby. You might also notice a dim branch of M20 extending away from the nebula's northern edge, almost like the stem of the pansy.

### Trifid Nebula

Messier was the first to set eyes upon M20 in 1764, although he described it only as a "cluster of stars." Today, we know that M20 is a glowing cloud of ionized hydrogen gas, a Hydrogen-II region like larger M8. Spanning about 25 light-years, M20 has enough mass to form several hundred suns over the course of its lifetime. In fact, the power that irradiates M20 comes from a young star cluster that has already formed. The brightest member of the cluster is the star HN 40.

M20 is more than just a cloud of ionized hydrogen. The dark rifts that give it its nickname are actually lanes of opaque dust, hiding whatever lies behind. There is also more dust found in that faint, northern extension of M20. Technically not part of the Hydrogen-II cloud, this portion of M20 is a blue reflection nebula, which is caused by dust particles that are illuminated by the light of nearby stars.

## M21: Open Cluster in Sagittarius

DISTANCE FROM EARTH: 4,200 light-years
FINDING FACTOR: **
"WOW!" FACTOR:    Binoculars: **      Small telescopes (3″ to 5″): **
                 Medium telescopes (6″ to 8″): **

### Where to Look

M21 can be found less than one degree to the northeast of M20. Both should just fit into the same field of a low-power eyepiece.

### What You'll See

**Through Binoculars.** M21 is a bright open cluster that is as impressive for its surroundings as it is for itself. Dozens of stars as bright as 8th magnitude call M21 home. Most binoculars will show the brightest of them, with the rest blending into a soft glow.

**Through a Telescope.** Through small instruments, M21 is easily resolved into about two dozen points of light highlighted by a string of stars near its center. Larger scopes increase the star count to about sixty, with several forming close pairs.

## M24: Star Cloud in Sagittarius

NICKNAME: Small Sagittarius Star Cloud
DISTANCE FROM EARTH: 10,000 to 16,000 light-years
FINDING FACTOR: *
"WOW!" FACTOR:    Binoculars: ***      Small telescopes (3″ to 5″): **
                 Medium telescopes (6″ to 8″): *

### Where to Look

M24, also known as the Small Sagittarius Star Cloud, can be seen with the naked eye from dark, rural locations. To zero in on it with binoculars or a finderscope, extend a line from the star Kaus Australis to Kaus Media in the Teapot, then continue the line about one and a half times that distance to the north, to the star Mu Sagittarii. From there, look for a small, kite-shaped pattern of stars to its north-northeast. See that soft glow surrounding the stars? That's M24.

### What You'll See

**Through Binoculars.** Overall, M24 paints a very pretty scene through small, wide-field binoculars. With even the slightest magnification, the whole area disintegrates into countless points of light of many magnitudes.

**Through a Telescope.** As pretty as the view through binoculars is, M24 is very disappointing through most telescopes, because of its size. Covering an area equal to three Full Moons stacked end to end, M24 is far too large to fit into a

### Small Sagittarius Star Cloud

For years, the identity of M24 was debated. Some astronomers believed it to be the same as NGC 6603, although this did not match Messier's original description from 1764. It wasn't until the 1960s that the late Kenneth Glyn Jones of the British Astronomical Association suggested that M24 was actually a larger region encompassing NGC 6603 known as the Small Sagittarius Star Cloud. Its visual appearance seemed to match Messier's words, and so a two-century-old mystery was finally solved.

M24's description as a "star cloud" is rather misleading, since it is not a single entity like a star cluster or nebula. There is a great deal of dust within the spiral arms of the Milky Way. This dust often occurs in clumps, which we call *reflection nebulae,* if illuminated by a nearby star, or *dark nebulae,* if not. In other areas of the Milky Way, the dust is not as densely packed, affording astronomers a clearer view of what lies beyond. M24 is one such gap or clearing in the dust clouds, allowing astronomers to peer more deeply toward the core of our galaxy. When we gaze toward M24, we are actually looking at the next spiral arm inward, giving us a taste of the Milky Way's inner workings.

single eyepiece field of view. Small rich-field telescopes are best for objects of this sort, but even these pale to the view through binoculars. If you don't own binoculars or a rich-field telescope, stick with your finderscope for the best view.

Although M24 leaves a lot to be desired through a telescope, a small open star cluster within its borders offers a challenging sight through 6-inch (15-cm) and larger instruments. Set just east-northeast of the eastern star in the "kite," NGC 6603 has a rich but faint glow. Don't expect to see many of the cluster's individual stars, however, as the brightest are only 11th magnitude. Most are as dim as the planet Pluto.

## M18: Open Cluster in Sagittarius

DISTANCE FROM EARTH: 4,900 light-years

FINDING FACTOR: ***

"WOW!" FACTOR:    Binoculars: *      Small telescopes (3″ to 5″): **
                 Medium telescopes (6″ to 8″): *

### Where to Look

From the "kite" embedded in M24, look to the north for a slender, south-pointing triangle of dim stars. M18 can be found just to the east of the triangle's apex.

### What You'll See

**Through Binoculars.** In some ways, M18 is like a small shack set in a neighborhood of mansions. While this area of Sagittarius is one of the finest regions for amateur astronomers, M18 is singled out only by its lack of presence. The brightest stars in this open cluster are set in a triangular pattern and may just be glimpsed in 10× glasses, with the rest blending into a faint haze.

**Through a Telescope.** Telescopes only do a little better with M18, expanding the star count to about eighteen. A star-filled backdrop contrasts nicely against the weak cluster. Smaller telescopes have the edge over larger instruments in this case, as they seem to differentiate better between the cluster and its surroundings.

## M17: Bright Nebula in Sagittarius

NICKNAME: Horseshoe, Omega, or Swan Nebula
DISTANCE FROM EARTH: 5,000 light-years
FINDING FACTOR: **
"WOW!" FACTOR:    Binoculars: ***        Small telescopes (3″ to 5″): ***
                 Medium telescopes (6″ to 8″): ****

### Where to Look

Return to that triangle used to find M18 and take aim at its brightest (eastern) star. M17 is just to its east and should be bright enough to be seen through a finderscope on fairly dark evenings.

### What You'll See

**Through Binoculars.** M17 is an easy target through nearly all binoculars.  With 7× binoculars you should be able to make out a conspicuous "bar" of light oriented more or less east-west within the glow of the nebula. Giant binoculars, or smaller glasses if supported steadily, may add a "reverse hook" curving off the western end of the bar.

**Through a Telescope.** Three-inch (7.5-cm) telescopes expand the view to reveal an intricate cloud in the shape of an extended number 2 or check mark (✓). In fact, it's the cloud's long, curved "neck" that gives rise to the nickname the Swan Nebula. The swan must be swimming, however, since we can't see its wings or feet. Increasing the aperture to 8 inches (20 cm), the Swan slowly disappears into a glowing semicircular arc of light that is shaped like a ghostly horseshoe or Greek capital letter omega.

M17 (Figure 7.12) is my favorite summertime nebula, and if it's up, it's one of the first objects I turn to every time I head outdoors. That diligence really pays off with M17, since new intricacies can be found with each return visit. My original notes of M17 made back in 1972 recall a rather bland view of the swan. Today, using the same 8-inch (20-cm) reflector, but now with experienced eyes, I can see some fantastic detail within the cloud, especially within and around the number 2's bottom "bar." Once again, a narrowband light-pollution reduction or Oxygen-III filter will greatly improve the view of M17, changing it from a vague outline to a distinct, intricate cloud of interstellar gas.

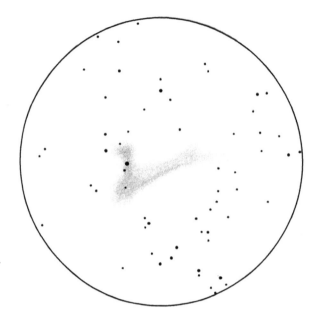

**Figure 7.12** *M17, the Swan Nebula, as drawn through the author's 8-inch reflector at 142× and aided by a contrast-enhancing narrowband light-pollution filter. South is up.*

### A Favorite Summertime Nebula

The Swiss astronomer Philippe Loys de Cheseaux discovered M17 in the spring of 1764, beating Charles Messier by just a few months. Messier was later to describe it as shaped like a spindle, a reasonable portrait for the brightest portion of the cloud. Observing through larger instruments, William Herschel was the first to liken its form to the Greek capital letter omega.

M17 is a glowing cloud of ionized hydrogen, much like M8 (the Lagoon Nebula) and M20 (the Trifid Nebula). There is enough "star stuff" in the Swan, or Omega, Nebula to create nearly a thousand stars the mass of our Sun. Over eons of time, M17 will beget an impressive open star cluster, which in turn will cause the remaining nebulosity to disperse.

To give a feel for the scale of what we are looking at, the bright bar that forms the straight bottom of the number 2 extends some 15 light-years (88 trillion miles or 142 trillion km)! But that's really only the tip of this celestial iceberg, as the full extent of M17 spans 40 light-years.

## M16: Bright Nebula and Open Cluster in Serpens

NICKNAME: Eagle Nebula
DISTANCE FROM EARTH: 7,000 light-years
FINDING FACTOR: **
"WOW!" FACTOR:    Binoculars: **        Small telescopes (3″ to 5″): **
                 Medium telescopes (6″ to 8″): ***

### Where to Look

From the triangle that led to M17 and M18, move one length farther to the north, where your finderscope should show a dim patch of fuzz, possibly with one or two very dim points shining within. That's M16.

### What You'll See

**Through Binoculars.** Although fairly obvious, M16 is not as satisfying an object through binoculars as M17 to its south. A 7× pair shows only a couple of the brightest suns that call M16 home, along with the dim flicker of fainter suns. Larger binoculars, 70 mm and greater in aperture, will show about a dozen stars.

**Through a Telescope.** Although M16 is best known by its nickname, the Eagle Nebula, most amateur telescopes show it as a scattered open cluster of about five dozen suns. But if you take a long-exposure photograph of M16, suddenly, from the depths of the star cluster, a great complex of beautiful nebulosity rises (Figure 1.3). I personally call it the Dr. Jekyll and Mr. Hyde Nebula. The Eagle Nebula itself, unseen by Messier and cataloged separately as IC 4703, is extremely dim and difficult to see visually. For years it went unseen by amateur astronomers, but thanks to the widespread use of contrast-enhancing narrowband and Oxygen-III light-pollution nebula filters, the Eagle can now be seen to soar where it had never been seen before. Because of the tenuous nature of the Eagle Nebula, however, you will need to wait for a dark, moonless night, even with a filter.

### Eagle Nebula

Studies show that M16 is one of the youngest star clusters known, perhaps no more than 5 million years old, affording astronomers an excellent laboratory for study. The hottest stars within M16, spectral type O, are seven times hotter than our Sun.

The Eagle Nebula, from which the stars of M16 formed, was the subject of perhaps the most famous photograph ever taken by the Hubble Space Telescope. In 1995, NASA released images that showed magnificent, inky black "pillars of creation" rising up against a backdrop of turquoise and orange. Likened to serpents heads, these columns are composed of cooled hydrogen and dust, the ingredients needed to form new stars. The Space Telescope Science Institute notes that as the pillars themselves are slowly eroded away, small globules of even denser gas buried within the pillars are uncovered. These globules are dubbed EGGs, an acronym for evaporating gaseous globules. Eventually most of these embryonic stars will evolve into mature stars.

## M23: Open Cluster in Sagittarius

DISTANCE FROM EARTH: 2,100 light-years
FINDING FACTOR: **
"WOW!" FACTOR:     Binoculars: **     Small telescopes (3″ to 5″): ***
                  Medium telescopes (6″ to 8″): ***

### Where to Look

To find M23, first relocate the star Mu Sagittarii, as detailed earlier when finding M24. From Mu, aim your telescope to the west-northwest until the star is just out of the finder's field of view. M23 should be centered in the field, looking like a hazy patch of starlight.

### What You'll See

**Through Binoculars.** M23 is a rich open cluster compressed into an area that appears as large as the Full Moon. Most binoculars resolve a few of its brightest points just breaking through a haze created by all of the other cluster suns that are too faint to be seen separately. An unrelated 6th-magnitude star lies just beyond the cluster's northwest corner.

**Through a Telescope.** M23 is one of the prettiest open star clusters in Sagittarius, but as with the view through binoculars, you may only see its brightest stars. The remaining cluster members blur together to create a grainy, background glow. Not surprisingly, given its wide span, low-power eyepieces are best for drinking it all in. The star pattern in M23 might remind you of a bird, or possibly a bat, in flight.

## M25: Open Cluster in Sagittarius

DISTANCE FROM EARTH: 2,000 light-years
FINDING FACTOR: **
"WOW!" FACTOR:    Binoculars: **      Small telescopes (3″ to 5″): **
                 Medium telescopes (6″ to 8″): **

### Where to Look

If you were able to find Mu Sagittarii and M23, then M25 should present a similar challenge. From Mu, nudge your finder toward the northeast, but keep Mu in the field. With a little imagination, you might see how Mu joins with three other, fainter stars to form a diamond pointing northeastward. Center on the eastern star in the diamond, then look just to its south for the soft glow of M25.

### What You'll See

**Through Binoculars.** M25 is a conspicuous cluster in nearly all binoculars. Although it's about the same apparent size as M23, M25 shows fewer individual stars through binoculars, although several cluster members are brighter than M23's brightest. In fact, the brightest star in M25 is U Sagittarii, a Cepheid variable star.

**Through a Telescope.** Most of the stars in M25 shine at 10th magnitude and below, so your impression will depend a lot on your telescope. A 3-inch (7.5-cm) instrument may only show perhaps half a dozen individual stars against a background glow, while a 6- or 8-inch (15- or 20-cm) telescope will quickly increase that number to several dozen. Take a close look at the cluster stars and you might notice a red ruby or two tucked in among its white diamonds. (These two reddish stars do not actually belong to M25 but are just chance line-of-sight affiliates.)

### M25

Discovered by Philippe de Cheseaux in 1745 or 1746, M25 spreads across some 23 light-years. M25 is unique in two respects. First, it is the only Messier object that is not in the *New General Catalog* but rather is cross-listed in the NGC's supplement, the *Index Catalogue* (IC). Normally, IC objects were discovered after the original NGC was published in 1888, but M25 had been known for more than 140 years when the NGC came out.

The second interesting fact about M25 is that it is one of only a few open clusters known to contain a Cepheid-type variable star. U Sagittarii cycles between magnitudes 6.3 and 7.1 over a period of 6 days 18 hours. See if you can follow its fluctuations over several nights by comparing it to other stars within the cluster. The fact that one of M25's stars has evolved into a Cepheid variable tells astronomers that M25 is relatively old as star clusters go, possibly close to 90 million years.

## M22: Globular Cluster in Sagittarius

DISTANCE FROM EARTH: 10,000 light-years
FINDING FACTOR: *
"WOW!" FACTOR:     Binoculars: **     Small telescopes (3″ to 5″): ***
                   Medium telescopes (6″ to 8″): ****

### Where to Look

Begin your quest for M22 at the star Kaus Borealis at the top of the Sagittarius Teapot. Scan half a finder field to the northeast, where you should see a small triangle of stars and, just beyond, the soft glow of M22.

### What You'll See

**Through Binoculars.** Under ideal conditions, M22 is visible to the unaided eye as a hazy spot northeast of Kaus Borealis. Binoculars that are 7× easily show a brighter core surrounded by a nebulous disk of light. High-power binoculars further enhance M22's appearance, but resolution of its stars is only possible through telescopes.

**Through a Telescope.** Small backyard instruments show M22 as a round, "grainy" disk with a mottled appearance. A 4-inch (10-cm) is probably the smallest scope that will begin to resolve some stars around M22's fringe, while an 8-inch (20-cm) is more than capable of displaying hordes of 11th-magnitude and fainter stars right across the cluster's core.

Take a slow, careful look at M22. Notice anything unusual about its shape? Most globular clusters are round, but not M22. A northeast-southwest bulge is noticeable through 3-inch (7.5-cm) telescopes and might even be seen in binoculars.

### M22

M22 (Figure 1.2) was the first globular cluster ever discovered, glimpsed by the little-known German astronomer Abraham Ihle in 1665. Were it not for its low altitude from the Northern Hemisphere, M22 might be the standard by which all other globulars are judged, rather than better known M13 in Hercules (see Summer Sky Window 2). In reality, however, M22 is not nearly as large as M13, containing less than half as many stars. But M22 is also considerably closer. M22 lies about 10,000 light-years away and has a real diameter of about 65 light-years, while M13 is about 25,000 light-years away and measures some 165 light-years across.

## M28: Globular Cluster in Sagittarius

DISTANCE FROM EARTH: 18,600 light-years

FINDING FACTOR: **

"WOW!" FACTOR:     Binoculars: *     Small telescopes (3″ to 5″): **
                   Medium telescopes (6″ to 8″): **

### Where to Look

M28 is easier to find than M22, but being smaller and fainter, it can be more difficult to actually see. Center on Kaus Borealis at the top of the Teapot and look through your telescope. Kaus Borealis and M28 are separated by about 1°, or two Full Moon diameters, and may just squeeze into the same field of your telescope's lowest-power eyepiece. If not, center on Kaus Borealis and nudge the view so that the star slides to the southeast. As it moves to the side, M28 should come into view from the opposite edge.

### What You'll See

**Through Binoculars.** Through binoculars, M28 is seen as a small smudge of faint light just northwest of Kaus Borealis. It is rather dim, especially compared to the more impressive M22, so it may take a little while to locate, especially if your view is affected by light pollution.

**Through a Telescope.** Telescopes also show M28 as a small, concentrated ball of light. Trying to resolve its stars proves to be a tough task through telescopes smaller than about 6 inches (15 cm) in aperture, since the brightest only shine around 12th magnitude. You might see a bit of "graininess" in M28, suggesting that stellar resolution is imminent. Since M28 is so small and concentrated, it can take magnification well. Your best view will probably come with 100× or more.

## M54: Globular Cluster in Sagittarius

DISTANCE FROM EARTH: 89,000 light-years

FINDING FACTOR: ***

"WOW!" FACTOR:     Binoculars: *     Small telescopes (3″ to 5″): **
                   Medium telescopes (6″ to 8″): **

### *M54: Extragalactic Globular?*

M54, discovered by Charles Messier in 1778, is the most distant globular cluster in the Messier catalog. In fact, studies released in 1994 show that M54 may actually be a member of another galaxy. That year it was announced that a previously unseen galaxy, called SagDEG or the Sagittarius Dwarf Elliptical Galaxy, had been discovered by R. Ibata, M. Irwin, and G. Gilmore. M54 is believed to be at about the same distance as this galaxy, and also appears to be moving in the same direction. The galaxy had remained unknown until 1994 because it is heavily obscured by interfering dust clouds within the Milky Way.

Messier missed nearby M69 and M70, not adding them to his catalog until 1780.

### Where to Look

M54 is the easternmost of three Messier globular clusters that lie along the bottom of the Sagittarius Teapot, between the stars Ascella and Kaus Australis. Aim toward Ascella, then keep an eye out for a triangle of dim stars about half a field to the west-southwest. Since M54 might be too faint to see directly through your finderscope, center the triangle in your telescope's low-power eyepiece and then shift your aim just to its north.

### What You'll See

**Through Binoculars.** M54 looks like a very dim, slightly out-of-focus star through most binoculars. Its concentrated core appears surrounded by a dim glow, but little additional detail will be visible.

**Through a Telescope.** Through 3- to 6-inch (7.5- to 15-cm) telescopes, M54 displays a perfectly round disk of light, with no indication of its true nature. Since the stars that form M54 shine no brighter than magnitude 12, an 8-inch (20-cm) aperture is needed to make out more than one or two weak points of light.

## M69: Globular Cluster in Sagittarius

DISTANCE FROM EARTH: 28,000 light-years
FINDING FACTOR: ***
"WOW!" FACTOR:     Binoculars: *     Small telescopes (3″ to 5″): **
Medium telescopes (6″ to 8″): **

### Where to Look

M69 is the second of three Messier globular clusters found along the bottom of the Sagittarius Teapot. Aim toward the star Kaus Australis, then look through your finder for a faint star just to its northeast. Continue to two other, brighter stars farther to the northeast. Aim at that pair, then look through your telescope with its lowest-power eyepiece. M69 lies just north of the western star in the pair.

### What You'll See

**Through Binoculars.** M69 is a challenging object through 50-mm binoculars and will be visible as just a faint, fuzzy "star."

**Through a Telescope.** M69 lies quite close to an 8th-magnitude field star, which can make seeing it difficult if your telescope optics are at all dirty (since dirt and grease on optical surfaces can scatter light). Once you know that you are aimed in the right direction, switch to a moderate magnification (75× to 100×) to isolate the cluster from the star. Most backyard telescopes will show M69 as an unresolved mass, with no hope of spotting any of its dim suns. It takes at least a 10-inch (25-cm) telescope to crack M69's stellar vault.

## M70: Globular Cluster in Sagittarius

DISTANCE FROM EARTH: 29,400 light-years
FINDING FACTOR:  ***
"WOW!" FACTOR:    Binoculars: *      Small telescopes (3″ to 5″): **
                 Medium telescopes (6″ to 8″): **

### Where to Look

M70 lies almost exactly halfway between the stars Kaus Australis and Ascella along the bottom of the Teapot. Trying to spot it will prove quite challenging, since there are no nearby stars bright enough to be seen through most finderscopes. Aim your telescope at the halfway point of the Teapot, then switch to your lowest-power eyepiece and start scanning slowly. M70 forms a triangle with two 8th-magnitude stars, one just to its west and another to its south.

### What You'll See

**Through Binoculars.** Visible faintly through 7 × 50 and 10 × 50 binoculars, M70 appears as little more than a slightly fuzzy star.

**Through a Telescope.** M70, like M54 and M69, challenges observers. Just trying to spot these three objects is a tough test. But even once you bump into M70, don't expect to see much detail. Its stars are so dim and tightly packed that it takes at least a 10-inch (25-cm) telescope to resolve any individuals. The rest of us will see M70 as a round, compact mass with a bright central core.

## M55: Globular Cluster in Sagittarius

DISTANCE FROM EARTH: 17,600 light-years
FINDING FACTOR:  ***
"WOW!" FACTOR:    Binoculars: **     Small telescopes (3″ to 5″): ***
                 Medium telescopes (6″ to 8″): ***

### Where to Look

Extend a line from the stars Nunki to Tau Sagittarii in the Teapot's handle. Following that line through your finderscope, you will first pass a small triangle of dim stars at about the halfway point to M55. Continue another finder field or so to the southeast to find M55, lying within a long, thin rectangle of 6th-magnitude stars.

One word of warning. Unlike other globular clusters, which appear highly concentrated, M55 is spread out more evenly. As a result, its surface brightness (or brightness per unit area) is deceptively low. So don't be surprised if finding M55 is a little more difficult than you might expect at first.

### What You'll See

**Through Binoculars.** Through binoculars, M55 appears as a dim ball of light. If you have sharp eyes and steadily held binoculars, you just might make out a bright point of light slightly off-center in M55. That lone star, shining at about 9th magnitude, is most likely a foreground object and not actually a member of the cluster.

**Through a Telescope.** Because its concentration is so low, M55 is easy to resolve into individual stars. It is one of the few globulars that actually look like a globular through small telescopes. Even a 3-inch (7.5-cm) instrument will partially resolve the group, although you'll need a good view to the south, away from interferences caused by light pollution and smog. A low-power 6- to 8-inch (15- to 20-cm) telescope shows many individual points of light sprinkled across the nucleus.

## M75: Globular Cluster in Sagittarius

DISTANCE FROM EARTH: 61,000 light-years
FINDING FACTOR: ****
"WOW!" FACTOR:    Binoculars: *      Small telescopes (3″ to 5″): **
                 Medium telescopes (6″ to 8″): **

### Where to Look

M75 is found in far eastern Sagittarius, not far from the constellation Capricornus. Although there are no nearby bright stars to act as finding beacons, there are a few celestial landmarks along the way that will help. To find M75, extend the line that forms the bottom of the Teapot, from Kaus Australis to Ascella, to the east. Partway, watch for a distinctive double star to the north of the line and, a little farther along, a cross-shaped pattern to its south. The line will end at a small, faint triangle of stars. Center this triangle in your telescope's low-power eyepiece, then shift just to the northeast to find the dim glow of M75.

### What You'll See

**Through Binoculars.** Rated at only magnitude 8.6, M75 is a tough catch with binoculars. Yes, it is possible but only by steadily supporting the binoculars against a wall, table, or preferably, on a tripod. It will look like a perfectly round, compact object accented by a brighter nucleus.

**Through a Telescope.** The compact structure of M75 shows up as a small, highly concentrated object through backyard telescopes. Unless you are viewing through a 10-inch (25-cm) telescope or larger, when a few faint stars might be spotted, you will most likely see just an ill-defined ball of fuzz. Even through 16- and 18-inch (40- and 45-cm) telescopes, M75 is only willing to show some individual stars around its periphery.

## Summer Sky Window 6

The southernmost star in the Summer Triangle, Altair, serves as a sighting beacon for Summer Sky Window 6 (Figure 7.13). Immersed in the Milky Way, this region of the summer sky is rich in stars, especially within the tiny constellation Scutum. Although the constellation pattern is drawn from faint naked-eye stars, the Scutum Star Cloud is a well-known feature of the summer sky, as is one of its denizens, M11.

### M11: Open Cluster in Scutum

NICKNAME: Wild Duck Cluster
DISTANCE FROM EARTH: 6,000 light-years
FINDING FACTOR: *
"WOW!" FACTOR:    Binoculars: ****    Small telescopes (3″ to 5″): ****
                 Medium telescopes (6″ to 8″): ****

#### Where to Look

To find M11, first look for the constellation of Aquila, marked by brilliant Altair. Trace out the Eagle's diamond-shaped body, then follow the curve of its tail-feather stars, Lambda and 12 Aquilae, which team with Eta Scuti to form a three-star arc that will point you toward M11.

#### What You'll See

**Through Binoculars.** Through binoculars, M11 looks more like an unresolved globular cluster than an open cluster. Except for a lone 8th-magnitude maverick shining off-center, its hundreds of stars all shine between 11th and 14th magnitude, so they cannot be resolved through most binoculars.

**Through a Telescope.** Here's one for your "favorites list." M11 (Figure 7.14) is a spectacular object, one of the richest, brightest open clusters visible from our planet. To its discoverer, Gottfried Kirch, M11 was "a small, obscure spot with a star shining through." That's exactly what you will see by viewing M11 through a 3- or 4-inch (7.5- to 10-cm) telescope. That lone star shines at 8th magnitude and is set amid the nebulous glow of fainter cluster stars. Larger instruments burst the group into an amazing swarm of hundreds of fainter stars that shine between 11th and 14th magnitude.

Collectively, M11's stars appear to form a blunt V pattern. To the famous eighteenth-century astronomer Admiral Smyth, they resembled a flock of wild

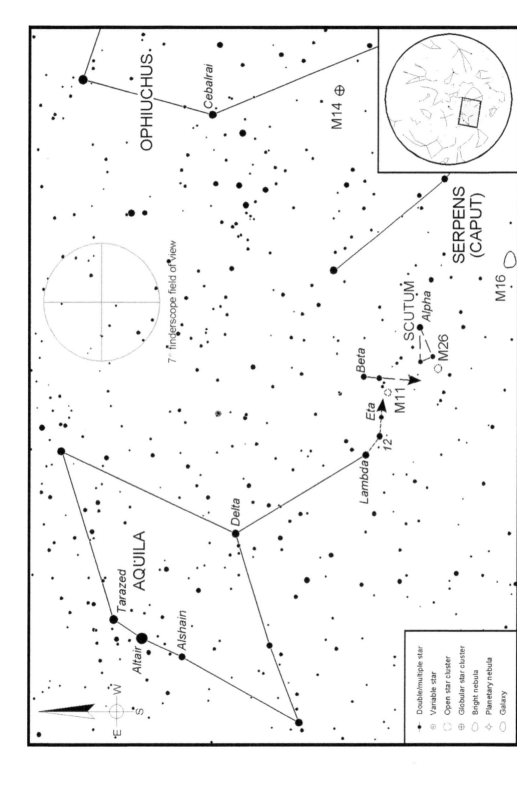

**Figure 7.13** *Summer Sky Window 6 includes portions of the constellations Aquila and Scutum.*

**Figure 7.14** *M11, the Wild Duck Cluster, is one of the sky's most densely packed open clusters. George Viscome used a 14.5-inch f/6 reflector to take this photograph, which is oriented with south toward the top.*

ducks, giving rise to M11's nickname, the Wild Duck Cluster. Because M11 is so compact, use a moderately high-power eyepiece for the best view. If you look carefully, you might be able to spot a few red suns among the throng of white points.

### *Wild Duck Cluster*

M11 was discovered in 1681 by the German astronomer Gottfried Kirch. More than a thousand stars belong to this open cluster, which spans roughly 24 light-years, making it the densest of its type among the M objects. Although most are young, hot type A and type F suns, the cluster's few type O and type B stars, which evolve more rapidly, have already left the Main Sequence and entered their red-giant phase. These studies allow astronomers to estimate that M11 is about 100 million years old.

## M26: Open Cluster in Scutum

DISTANCE FROM EARTH: 5,000 light-years
FINDING FACTOR:  **
"WOW!" FACTOR:   Binoculars: *       Small telescopes (3″ to 5″):  **
                Medium telescopes (6″ to 8″):  **

### Where to Look

Follow the steps to M11 given above, then continue a little farther west, to a pair of fairly bright stars (bright as seen through binoculars and finderscopes, that is). Place them in the northeast corner of your finder field, then look to the southwest for a small right triangle of stars. The star Alpha Scuti is the brightest star in the triangle. M26 lies just to the east of the triangle's hypotenuse. The cluster should be visible through 6 × 30 and larger finderscopes.

### What You'll See

**Through Binoculars.** M26 is a small, condensed open cluster made up of about thirty stars, none of which shines brighter than 10th magnitude. Though they are not bright enough to be seen through most binoculars, their light combines to produce a misty glow spanning about a quarter of a degree.

**Through a Telescope.** M26 is nowhere as visually stunning as M11, but even though it can't compare to its neighbor, it is still a pleasant enough star cluster through telescopes. Use a low-power eyepiece (around 50×), since higher magnifications tend to dilute the clustering effect against the rich surroundings.

> *M26*
>
> Discovered by Charles Messier in 1764, M26 is positioned within the Scutum Star Cloud, one of the richest regions of the Milky Way. In all, ninety or so stars inhabit M26, which has an estimated age of 89 million years and spans 22 light-years.

## Summer Sky Window 7

Farther north along the Milky Way, Summer Sky Window 7 (Figure 7.15) holds a wealth of buried treasures for stargazers, with a pair of showpiece planetary nebulae, as well as two of the summer's finest binary stars, a team of globular clusters, an open cluster, and a coathanger. A coathanger?!

### Albireo: Binary Star in Cygnus

DISTANCE FROM EARTH: 385 light-years
FINDING FACTOR: *
"WOW!" FACTOR:     Binoculars: **     Small telescopes (3″ to 5″): ****
                   Medium telescopes (6″ to 8″): ****

### Where to Look

Albireo, also known as Beta Cygni, marks the beak or head of Cygnus the Swan, one of the summer's most recognizable constellations, or if you prefer, the base of the Northern Cross.

### What You'll See

**Through Binoculars.** If you hold your binoculars steady, you just might make out that Albireo is not a single star at all but rather a pair of distant suns set

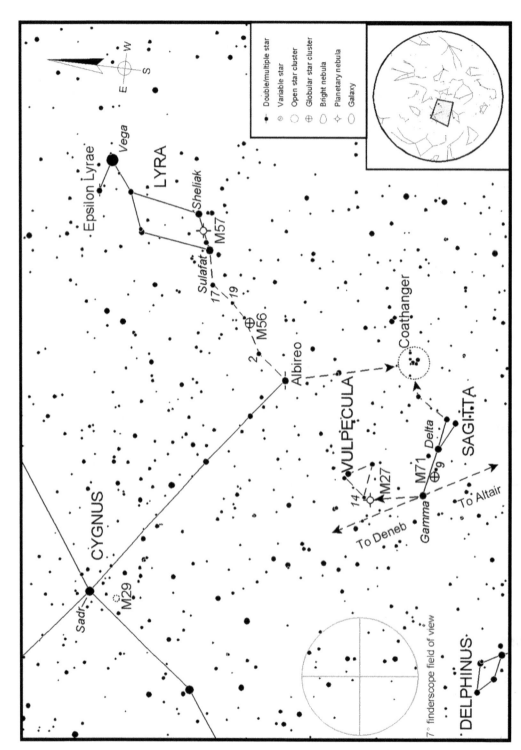

**Figure 7.15** *Summer Sky Window 7 includes the famous Ring Nebula as well as several other enjoyable targets.*

against a glorious Milky Way field. These two stars are famous for their outstanding color contrast, with brighter Albireo A a yellow, 3rd-magnitude star and 5th-magnitude Albireo B bluish in color.

**Through a Telescope.** Telescopes leave little doubt that Albireo is one of summer's most colorful stellar jewels. Albireo A is a gleaming 3rd-magnitude gem that shines with a golden radiance, while its companion, Albireo B, is a type B star that shines like a dazzling azure blue sapphire. These colors appear even more vivid if you first softly defocus the stellar images.

### Albireo

When the famous double star observer F. G. W. Struve first examined Albireo in 1832, he measured the apparent distance to the blue companion sun from Albireo A as 34 arc-seconds. Even though the stars have changed little in separation since, most binary-star authorities believe that the stars form a true binary system. Albireo B is estimated to be at least 4,400 astronomical units from Albireo A. At that tremendous distance, its period of revolution must be on the order of 100,000 years.

## Epsilon Lyrae: Multiple Star in Lyra

NICKNAME: Double-Double
DISTANCE FROM EARTH: 190 light-years
FINDING FACTOR: *
"WOW!" FACTOR:    Binoculars: **        Small telescopes (3″ to 5″): ***
Medium telescopes (6″ to 8″): ***

### Where to Look

Epsilon Lyrae, better known as the Double-Double, a quadruple star system, may be faintly glimpsed with the unaided eye near the dazzling star Vega, the brightest of three stars that form the Summer Triangle. Light pollution or haze easily obscure Epsilon, but even then, it should be simple enough to find through a finderscope or binoculars by centering on Vega and then looking just to the northeast.

If you look carefully with your naked eye, you just might be able to see that Epsilon is a double star. It's a tough test but a great way to check your eyes' visual acuity.

### What You'll See

**Through Binoculars.** Any binoculars will easily show that Epsilon is made up of, at the very least, two stars. The northern star is identified Epsilon-1, while the other is called Epsilon-2.

**Through a Telescope.** If you own at least a 3-inch (7.5-cm) telescope, try viewing the Epsilon twins at about 100×. If the air is steady and the seeing is good, each should resolve into a tight pair. Epsilon-1 is comprised of a

6th-magnitude secondary star and a 5th-magnitude primary star separated by 2.6 arc-seconds, with the companion set to the brighter star's north. The southern pair, Epsilon-2, is a bit tighter and therefore a little more difficult to resolve. The stars in this second pair are isolated from each other by 2.3 arc-seconds, with the 5.5-magnitude secondary star to the east of the 5.2-magnitude primary.

Each star is labeled in Figure 7.16 to make it clearer which is which, although note that the drawing is oriented with south at the top, as it would appear through an inverting telescope. The northern pair, Epsilon-1, can be further identified as Epsilon-1A and -1B, while the southern pair is identified as Epsilon-2A and -2B.

### Double-Double

Epsilon-1A and -1B orbit around a common center of gravity once every 1,100 years or so and lie about 150 astronomical units apart. Stars 2A and 2B are also separated by about 150 A.U. but only take about 585 years to orbit each other because of their greater masses.

Both pairs also orbit each other, though very slowly. The two pairs are separated by about 0.2 light-years (13,000 A.U.). At that great distance, they must take half a million years to complete one orbit around their common center of mass, which is located somewhere in the empty space between the pairs.

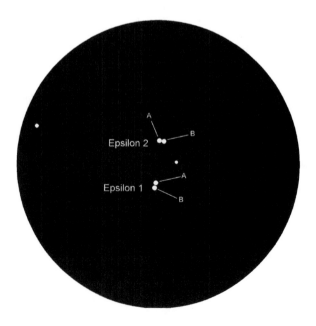

**Figure 7.16** *The components of the quadruple star Epsilon Lyrae, better known as the Double-Double, are labeled here for easy identification.*

## M57: Planetary Nebula in Lyra

NICKNAME: Ring Nebula
DISTANCE FROM EARTH: 2,300 light-years
FINDING FACTOR: *
"WOW!" FACTOR:    Binoculars: *        Small telescopes (3″ to 5″): ***
                  Medium telescopes (6″ to 8″): ****

### Where to Look

Although it is quite dim through small telescopes, M57 is easy to zero in on thanks to its location along the bottom of the Lyre's rectangular frame. Center your aim along the bottom of the Lyre's frame halfway between the star Sulafat marking the eastern corner and Sheliak at the western corner. M57 lies almost exactly between the two. Even at low power, your telescope should be able to distinguish the grayish disk of M57 from a star, but for a better view use a medium power after the nebula has been located.

### What You'll See

**Through Binoculars.** While it appears little more than a very dim star, M57 is indeed visible in binoculars. Admittedly, its famous smoke-ring shape requires more magnification than most binoculars can provide, but even through light-polluted suburban skies, I have seen it through 7 × 35 glasses. Give it a go!

**Through a Telescope.** Since its light is concentrated within a small disk, M57 (Figure 7.17) should be visible even from light-polluted cities if you look carefully for it. At first, with low power, it may look like little more than a fuzzy star, but by using averted vision and a reasonably high power (100× to 150×), the ring itself should become clear. In fact, M57's "doughnut" appearance can be seen by sharp-eyed observers viewing through telescopes as small as 2 inches (5 cm) across at about 100×. Though the Ring seems smooth and evenly bright in smaller instruments, 8-inch (20-cm) and larger telescopes will uncover some bright and dark irregular patches around its surface.

### Ring Nebula

Discovered by Antoine Darquier de Pellepoix in 1779, the Ring Nebula is the sky's most famous example of a planetary nebula. For years it was assumed that the distinctive shape of M57 was an illusion, that we were actually looking at an expanding spherical envelope of material ejected from its central star, and that the material appeared more densely packed around the edges only because of our perspective. Recent studies, however, suggest that the Ring is just that, a bright torus of glowing gases surrounding the star. Others believe that we are looking down a tunnel formed by the expanding material. The Ring measures about 0.9 light-years (5.5 trillion miles or 8.8 trillion km) across.

Although M57 looks grayish through backyard telescopes, many planetary nebulae appear bluish or greenish, testifying to the fact that they contain ionized oxygen, which is known to glow with a turquoise hue under certain conditions. Most planetary nebulae appear notoriously small, requiring high magnification to be seen.

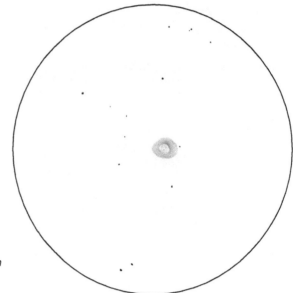

**Figure 7.17** *M57, the Ring Nebula, is perhaps the sky's most photographed planetary nebula. South is up in this drawing made through the author's 8-inch reflector at 120×.*

One of the great challenges facing deep-sky observers is trying to glimpse the central star that created the Ring so many millennia ago. Already faint to begin with, the central star's visibility is further confounded by the overwhelming brightness of the Ring itself, which washes it out. As a result, a 12-inch (30-cm) telescope is probably the smallest telescope capable of uncovering it, and then only under ideal conditions.

## M56: Globular Cluster in Lyra

DISTANCE FROM EARTH: 31,500 light-years
FINDING FACTOR: **
"WOW!" FACTOR:     Binoculars: *     Small telescopes (3″ to 5″): **
                   Medium telescopes (6″ to 8″): **

### Where to Look

M56 is settled about halfway in between the stars Albireo in Cygnus and Sulafat in Lyra. To find it, begin at Albireo. Looking through your finderscope or binoculars, move toward the northwest, pausing at the star 2 Cygni. From here, head a little west-northwest, toward the base of Lyra, but pause at a dim star directly south of a small triangle of dimmer stars. (If you were to continue toward Lyra, the trail would be completed by the stars 19 Lyrae and 17 Lyrae, as shown on Figure 7.15.) Place that star in your telescope's field of view and you should see M56 just to the southeast, in the same field of view.

### What You'll See

**Through Binoculars.** M56 doesn't put on much of a show through binoculars, looking like no more than a fuzzy 8th-magnitude smudge of light. Though unimpressive by itself, M56 combines with its surroundings to create a very pretty scene, which goes to prove that sometimes you are as good as the company you keep.

**Through a Telescope.** M56 is a tough nut to crack even through many amateur telescopes. Depending on the quality of its optics, a 6-inch (15-cm) telescope can show some of the cluster's stars, giving M56 what some observers describe as a "grainy" appearance. An 8-inch (20-cm) increases resolution, although an experienced eye and moderately high magnification are still required. The rest of us will have to settle for a round, ethereal glow. Unlike most globular clusters, M56 lacks a brighter central core.

## M71: Globular Cluster in Sagitta

DISTANCE FROM EARTH:  12,700 light-years
FINDING FACTOR:  ***
"WOW!" FACTOR:    Binoculars: *      Small telescopes (3″ to 5″):  **
                 Medium telescopes (6″ to 8″):  **

### Where to Look

Aim your finderscope toward the four stars that frame the small constellation of Sagitta the Arrow, settled about a third of the way from Altair to Deneb along the eastern side of the Summer Triangle. The stars are faint, so you may only be able to spot them through your finderscope or binoculars. Put the star Gamma Sagittae on the eastern edge of the field and Delta Sagittae toward the west. Just below and to the west of center, look for a dim star labeled as 9 Sagittae. Aim toward that star and have a look through your telescope, using your low-power eyepiece. M71, though dim, is found just to the star's east-northeast, in the same field of view.

### What You'll See

**Through Binoculars.** Binoculars only show M71 as a very dim splotch of grayish light against a star-filled backdrop.

**Through a Telescope.** A 3- to 4-inch (7.5- to 10-cm) telescope will show M71 as only a faint, ill-defined patch with few distinguishing characteristics. The stars that make up M71 (Figure 7.18) are well below the magnitude threshold for small and medium apertures. Indeed, just spotting M71 at first will probably not be a simple task. It appears rather small and faint, blending easily into the starry background without notice. You may need dark skies, patience, and a good dose of averted vision to make out its feeble glow.

## M27: Planetary Nebula in Vulpecula

NICKNAME:  Dumbbell Nebula
DISTANCE FROM EARTH:  1,250 light-years
FINDING FACTOR:  **
"WOW!" FACTOR:    Binoculars:  **      Small telescopes (3″ to 5″):  ***
                 Medium telescopes (6″ to 8″):  ****

**Figure 7.18** *M71, globular cluster in Sagitta. South is up. Photograph by George Viscome through a 14.5-inch f/6 reflector.*

### Where to Look

While spotting the outline of Vulpecula, the little-known constellation of the Fox, is nearly impossible, locating M27 is a bit easier. First, find the small constellation of Sagitta the Arrow. Gamma Sagittae, the eastern tip of Sagitta, lies a third of the way along a line that runs from Altair to Deneb in the Summer Triangle. From Gamma Sagittae, turn due north, to a triangle of faint stars. Look for M27 as a smudge of grayish light to the southeast of the star 14 Vulpeculae, the easternmost star in the triangle. Both easily fit into the same low-power eyepiece field.

### What You'll See

**Through Binoculars.** Equal to an 8th-magnitude star in brightness, M27 shows through 7× glasses as a hazy, rectangular smudge.

**Through a Telescope.** Famous as one of the brightest planetary nebulae in the sky, M27 is easily seen through even the smallest telescopes as a rectangular patch of light floating in a star-studded field. Unlike most planetaries, which look quite small through amateur telescopes, M27 is big. As a result, it is easy to identify at low power, although you will still want to switch to about 100× for the best view. If you have one, use a narrowband or Oxygen-III light-pollution filter as well.

M27 (Figure 7.19) is better known by its nickname the Dumbbell Nebula for its shape. Some see it as two lobes of light (the dumbbell's weights) con-

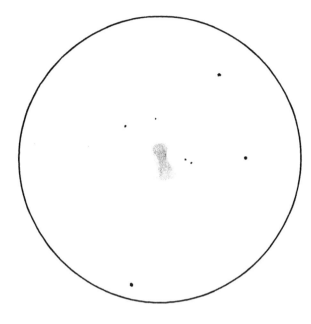

**Figure 7.19** *M27, the Dumbbell Nebula, in Vulpecula, as recorded through the author's 8-inch reflector at 55×, using a contrast-enhancing narrowband light-pollution filter. South is up.*

nected by a thin shaft of light (its bar). Others interpret the shape as an hourglass, bow tie, or even an apple core. The star that created the Dumbbell can only be seen faintly through 10-inch (25-cm) and larger backyard telescopes.

### Dumbbell Nebula

When Charles Messier discovered M27 on July 12, 1764, he could not have imagined what he had just stumbled upon. The first planetary nebula ever found, M27 is the outward-racing shell of gas expelled by a star similar in mass to the Sun but much farther along in its life. Astronomers believe that the expanding shell of gas was probably expelled some 3,000 to 4,000 years ago, and it continues to diffuse slowly.

More specifically, M27 is called a bipolar planetary nebula, as most of the material seems to be expanding from the star in two directions, while being blocked in other directions. Despite our ever-increasing knowledge of the universe, the mechanism that creates bipolar planetary nebulae is not well understood.

## Collinder 399: Asterism in Vulpecula

NICKNAMES: Coathanger or Brocchi's Cluster
DISTANCE FROM EARTH: Between 220 and 1,100 light-years
FINDING FACTOR: *
"WOW!" FACTOR:    Binoculars: ****     Small telescopes (3″ to 5″): ***
                 Medium telescopes (6″ to 8″): **

### Where to Look

The Coathanger, cataloged officially as Collinder 399, is visible to the naked eye on clear, dark nights, hanging dimly in the lane of the Milky Way. Finding it with binoculars is simply a matter of dropping about 1.5 fields due south from the star Albireo in Cygnus. It is also easy to spot just to the northwest of the constellation Sagitta.

### What You'll See

**Through Binoculars.** Without a doubt, the Coathanger is one of the nicest binocular deep-sky objects in the summer sky. Binoculars easily reveal how this pattern came to be known as the heavenly Coathanger, even though here in the Northern Hemisphere, the Coathanger is upside down. Six stars aligned in a straight line form the Coathanger's cross bar, while four other points of light curve away to create the hook. Binoculars that are 7× to 10× deliver the most pleasing view, since they will show the Coathanger suspended within a magnificent field filled with stardust.

**Through a Telescope.** You know the old saying "you can't see the forest for the trees." In this case, most telescopes can't show the Coathanger for all the stars. Tip to tip, the Coathanger measures 2° wide by 1° high, more than most telescopes can squeeze into their fields of view. As such, even though the stars glisten brightly through telescopes, the overall effect is lost.

### Coathanger

Discovered as far back as A.D. 964 by Al Sufi, the Coathanger was skipped entirely by Messier and Herschel, probably because of its large apparent size. For years it was assumed that the Coathanger was a true open star cluster, and in fact it was given a second nickname, Brocchi's Cluster, for the American celestial cartographer who drew many of the maps used by the American Association of Variable Star Observers (AAVSO). It was even cataloged as entry number 399 in the 1931 catalog of scattered open star clusters compiled by the Swedish astronomer Per Collinder. Much more recent data gathered by the European Space Agency's *Hipparchos* satellite, however, suggest this is not a true cluster at all but rather just a chance alignment of random stars that lie anywhere from 220 to 1,100 light-years away. Regardless, the question of whether the Coathanger is a true star cluster or a celestial fraud certainly does not detract from its appeal.

## Summer Sky Window 8

The final summer sky window (Figure 7.20) completes the detailed picture of Cygnus the Swan, a rich area that is wonderful for a casual scan with binoculars or a telescope. Start by finding the selected objects here, but be sure to allot enough time just to "star surf" through the area afterward, especially if you are fortunate enough to be out under dark skies, far from civilization.

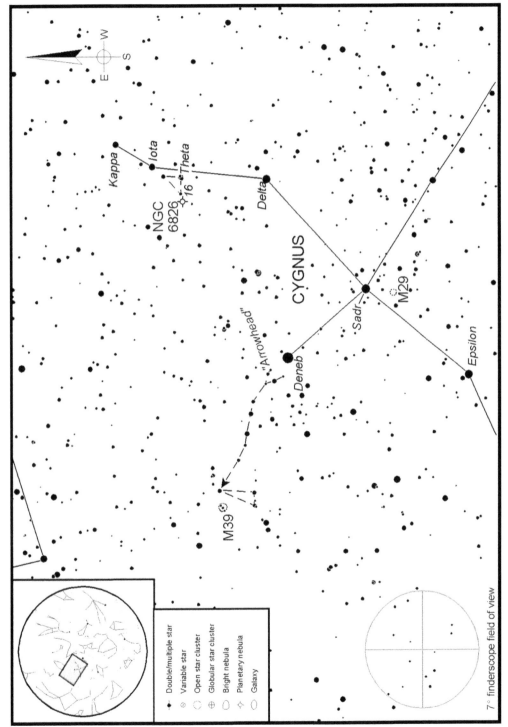

**Figure 7.20** *Summer Sky Window 8 covers northern Cygnus the Swan.*

## M29: Open Cluster in Cygnus

DISTANCE FROM EARTH: 4,000 light-years
FINDING FACTOR:  *
"WOW!" FACTOR:    Binoculars: **      Small telescopes (3″ to 5″): **
                 Medium telescopes (6″ to 8″): **

### Where to Look

M29 is located in the same finderscope field as Sadr, the center star in Cygnus the Swan. You should see two faint stars to Sadr's south, then the small, hazy splotch of M29 just beyond.

### What You'll See

**Through Binoculars.** M29 is one of the poorer open clusters in the Messier catalog. Binoculars will show a small, rectangular patch of light, with perhaps two or three faint points buried within.

**Through a Telescope.** Although more than a dozen of its stars are visible through small amateur telescopes, M29 is so sparsely packed that it can be easily mistaken for just a rich background star field rather than an open cluster. Look for a short dipper pattern that is reminiscent of the much brighter Pleiades in the winter sky. Four of the cluster's brightest stars form the rectangular bowl, while a fifth star can be imagined as a stubby handle. Be sure to use your lowest power, as higher magnifications only dilute the clustering effect.

## NGC 6826: Planetary Nebula in Cygnus

NICKNAME:  Blinking Planetary
DISTANCE FROM EARTH: 2,200 light-years
FINDING FACTOR:  **
"WOW!" FACTOR:    Binoculars: *      Small telescopes (3″ to 5″): **
                 Medium telescopes (6″ to 8″): **

### Where to Look

From the star Sadr at the heart of Cygnus, follow the Swan's westward-extending wing to its tip at Iota and Kappa Cygni. Trace a line from Kappa through Iota toward the southeast, where you will find a small triangle formed from Theta Cygni, 16 Cygni (a nice double star in its own right), and a third, fainter star. NGC 6826 is found just east of 16 at the triangle's eastern point. Both fit into the same low-power eyepiece field, although you will probably need to increase magnification to identify the planetary nebula among all the field stars.

### What You'll See

**Through Binoculars.** The Blinking Planetary shines at about 9th magnitude, which is bright enough to be seen through 50-mm binoculars used under dark skies. But which "star" is actually the nebula? That's the problem facing observers. It's the first point of light due east of 16 Cygni and should look bluish. Averted vision will make NGC 6826 look a little fuzzy, while stars will stay pinpoint sharp.

**Through a Telescope.** NGC 6826 is nicknamed the Blinking Planetary because of the disappearing act it performs through amateur telescopes. When viewed out of the corner of your eye, NGC 6826 shows as a small but distinctive bluish or greenish disk. Its 10th-magnitude central star is also easy to see through telescopes as small as 2 inches (5 cm) across. But when you give in to temptation and try to look directly at the planetary, it disappears! Of course, the nebula didn't actually vanish. Instead, the light from the central star overpowers the nebula when both are viewed with the less-sensitive central part of your eye's retina. Only when their combined light falls along the retina's more sensitive edges can the nebula and star both be distinguished separately.

## M39: Open Cluster in Cygnus

DISTANCE FROM EARTH: 825 light-years
FINDING FACTOR: **
"WOW!" FACTOR:     Binoculars: **     Small telescopes (3″ to 5″): ***
                  Medium telescopes (6″ to 8″): ***

### Where to Look

Although M39 is quite bright, its out-of-the-way location in northeastern Cygnus makes it tough to spot at first. The easiest way to get there is to cast off from Deneb. Look for a small arrowhead of stars just to its northeast, which points toward a string of six faint stars that continues on a northeastern track. Aim at the sixth (eastern) star in the line, then look for M39 as a small patch of starlight a little farther to the east still. That sixth star also marks the northern point in a slender triangle, which might also help point the way to the cluster.

### What You'll See

**Through Binoculars.** Binoculars reveal about two dozen faint stars in M39 set in a triangular pattern. Using 10× or higher binoculars, you might even get the sense that the stars in M39 are dangling in front of the fainter field stars. It's a neat illusion!

### M39

M39 holds a footnote in history as one of the first deep-sky objects ever discovered, when the ancient Greek philosopher and scientist Aristotle spotted it in 325 B.C. Messier independently discovered his catalog's thirty-ninth entry in 1764.

In all, thirty stars scattered across 7 light-years belong to M39, with most bright enough to be seen through amateur telescopes. M39 is one of the closer open clusters and is estimated to be between 230 and 300 million years old.

**Through a Telescope.** Famous as a bright, loose congregation of stars and covering an area of sky as large as the Full Moon, M39 is best appreciated at very low powers. It's an ideal object for the 3.1- and 4-inch (8- and 10-cm) "short-tube" refractors that are so popular today. At 20×, these telescopes transform M39 into about two dozen blue-white stellar sapphires poured into a triangular pattern, but be forewarned that the cluster may lose some of its appeal when viewed with longer focal-length telescopes or at higher power.

# 8

# Autumn Sky Windows

Just as the summer sky was dominated by a large triangular pattern, the autumn sky (Figure 8.1) is centered on another geometric shape—a square. Flying high in the southern sky are the four stars of the Great Square of Pegasus the Flying Horse. Trying to imagine a horse among the stars of Pegasus, let alone one that flies, is a difficult task. Perhaps you can see him flying upside down, with the square representing the body. The horse's neck and head curve from Markab to Enif, its front legs extend above the square, but the tail end of the horse is nowhere to be found! Since autumn is traditionally associated with baseball and the World Series, it might be easier to see a baseball diamond among the stars of Pegasus. Take Scheat as home plate, Alpheratz as first base, Algenib as second, and Markab as third. A faint star in between second and third (not shown on the map) might be the shortstop, while another faint, unplotted star in the center of the square represents the pitcher's mound. And just look at all the fans scattered all across the sky!

Technically, the star Alpheratz does not belong to Pegasus but instead to the neighboring constellation Andromeda the Princess. Andromeda is formed by two lines of stars framing her body. Here we also find the most distant object visible to the unaided eye—the Andromeda Galaxy.

Andromeda is a member of the royal family of autumn that also includes King Cepheus and Queen Cassiopeia. Cassiopeia is formed by five stars set in a stretched W pattern. The stars of Cepheus, all much more difficult to pick out than those of Cassiopeia's W, frame what looks like a simple drawing of a house, with the peak of the roof pointing roughly toward Polaris.

According to ancient Greek legend, Cepheus and Cassiopeia ruled over ancient Ethiopia. Cassiopeia was well known for two things: her great beauty and her unabashed boastfulness. One day, she bragged of being fairer than

# *Autumn sky*

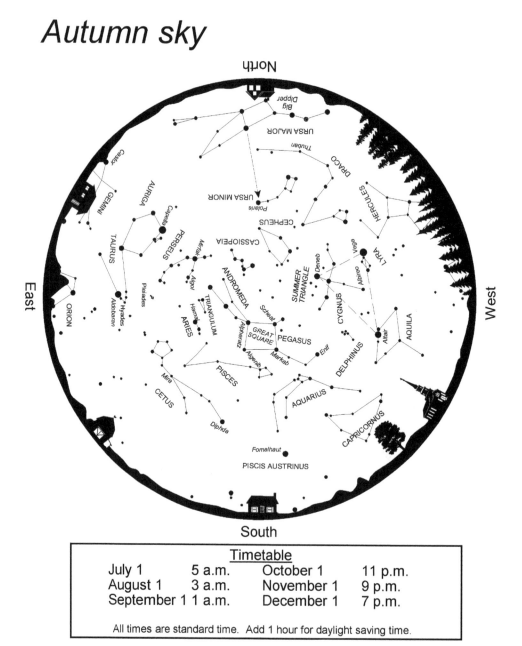

**Figure 8.1** *The autumn evening sky.*

the sea nymphs, who were well known for their exquisite beauty. The sea nymphs overheard this boasting and complained to their father, Poseidon, the king of the seas. Poseidon became so infuriated that he created a flood and Cetus the Sea Monster. Cetus was sent to Ethiopia to devour the citizens of the land. King Cepheus was told that his people could be saved only if he were to sacrifice his daughter Andromeda to the sea monster. As king, he had no alternative but to lead his daughter to the water's edge and chain her to a rock. There she was left to the mercy of Cetus.

Just then our hero, Perseus, appeared on the scene. Perseus, the son of Zeus, had been ordered to kill a gorgon called Medusa. Medusa was a very ugly creature whose hair was made of snakes. In fact, she was so ugly that anyone who looked at her face would instantly turn to stone. In order for Perseus not to turn to stone, he was given a highly polished shield and told not to look directly at Medusa, but only at her reflection. He was also given a helmet that made him invisible and a pair of winged sandals that allowed him to fly. With these, Perseus was able to sneak up on and quickly decapitate the horrible creature. He then put her head in a leather bag. Some of Medusa's blood fell into the water to create Pegasus the Flying Horse. Perseus climbed on the back of Pegasus and flew away.

Meanwhile, back at the seashore, things were looking pretty grim for Andromeda. Perseus heard her cries for help and swooped down to rescue her. Telling Andromeda to close her eyes, he pulled Medusa's head from the bag and dangled it in front of Cetus, who instantly changed to stone. Perseus had rescued Andromeda, and they fell in love. They climbed on the back of Pegasus and flew off into the sunset.

In our autumn sky, Perseus can be found towering protectively over Andromeda, holding a sword in one hand and the awful head of Medusa in the other. An interesting star in Medusa's head is Algol, which means "demon." This star, marking the winking eye of Medusa, varies in brightness every three days. These variations can be easily observed with the naked eye.

Cetus the Sea Monster (today, usually referred to as a whale) also belongs to the autumn sky. Found looming below the Great Square, the many faint stars of Cetus make the Sea Monster a difficult fish to catch from urban and suburban skies. Its brightest star, Diphda, rides so low in the sky that it may be blocked from view by trees or other obstructions. Even more interesting is the variable star Mira, which means "wonderful." At times, Mira will shine about as brightly as Polaris, while at others, it is nowhere to be found. It takes Mira more than a year to complete one full cycle, from bright to dim, back to bright again.

Skimming autumn's southern horizon is Fomalhaut, brightest star in the otherwise-obscure constellation Piscis Austrinus, the Southern Fish. Above Fomalhaut are many dim stars that our ancestors assigned to Aquarius the Water Bearer, the figure of a man carrying a large jar on his back. In most depictions, he is shown flooding the sky, with the water flowing through Capricornus, Piscis Austrinus, Cetus, and Pisces.

Like Aquarius, Pisces, the Fishes, is a zodiacal constellation. Except for that distinction, little attention would be paid to Pisces because its stars are so faint. The constellation is usually drawn as two fishes joined at the tails by a long rope.

East of Pisces along the ecliptic lies the small constellation Aries the Ram. It contains only one bright star—Hamal—which combines with two fainter suns to form the head of the animal. Aries was the legendary Ram of the Golden Fleece. When the Ram died, its valuable golden fleece was placed in a sacred oak tree that was guarded by a hydra, a ferocious, multiheaded dragon. Wanting to return the valuable golden fleece to its home country of Thessaly, Jason, heir to the country's throne, set out to find it. To help him in his quest, Jason gathered some of the mightiest warriors the world ever knew, among them Hercules, Orpheus, Castor, and Pollux. The ship they sailed on was named *Argo*, and its crew became immortalized as the Argonauts. Together, Jason and the Argonauts fell into some high adventures but never did find the fleece.

Though we do not find Jason in the stars, many of the Argonauts as well as portions of their ship are scattered across all four seasonal skies.

Tucked between Aries and Andromeda, we find the small constellation Triangulum the Triangle. Finally, a constellation that looks like what it represents!

Figure 8.2 holds the key to the autumn sky windows. The map, a duplicate of Figure 8.1, shows seven highlighted areas that are discussed in detail below. Each of those windows shows an enlarged area of the sky plotting fainter stars and selected deep-sky objects. Compare the two maps, decide which area you wish to explore tonight, and then head outside to find that region. Once the naked-eye stars and constellations have been located, switch to the close-up sky window found later in this chapter and, using your finderscope or binoculars, begin your journey into the depths of the universe.

Each sky window includes a miniature all-sky chart to show its position relative to the rest of the night sky. To help readers better appreciate the scale of things, a crosshair frame is also included in each window to show the typical field of a 6 × 30 finderscope (about 7°).

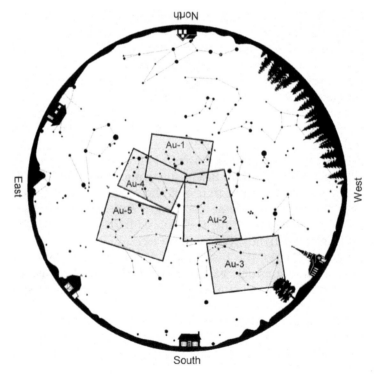

**Figure 8.2** *Autumn key map, showing each of the autumn sky windows to come.*

## Autumn Sky Window 1

Autumn Sky Window 1 (Figure 8.3) includes much of the constellations Cassiopeia and Cepheus, two circumpolar constellations for many of us in the Northern Hemisphere. As a result, while this region is visible highest in the sky during the autumn months, many of the objects within can be seen throughout the year if you have a good view to the north.

### M52: Open Cluster in Cassiopeia

DISTANCE FROM EARTH: 5,000 light-years
FINDING FACTOR:  *
"WOW!" FACTOR:    Binoculars:  ***     Small telescopes (3″ to 5″):  ***
Medium telescopes (6″ to 8″):  ***

#### Where to Look

Begin your quest for M52 at the W-shaped constellation Cassiopeia. Extend a line from Schedar to Caph, the westernmost stars in the pattern, an equal distance to the northwest, where you should spot a narrow diamond pattern. M52 lies just to the south of the diamond, and should be visible through your finderscope.

#### What You'll See

**Through Binoculars.** M52 is a beautiful sight through binoculars. Several individual suns are resolvable in binoculars, but the rest blur into a nebulous mass, which is exactly how Messier himself described this group. While in the area, be sure to scan the entire region slowly with your binoculars. The constellation Cassiopeia lies in one of the richest regions of the autumn Milky Way and is perfect for a casual scan through binoculars. Pull out a chaise lounge and enjoy the view!

**Through a Telescope.** M52 is a very nice sight through backyard telescopes. A 3- to 4.5-inch (7.5- to 114-cm) telescope shows a triangular swarm of faint suns highlighted by a distinctly brighter, orange star slightly off center. These are seen against the soft glow of other cluster members that are too faint to be resolved individually.

### Delta Cephei: Double Star/Variable Star in Cepheus

DISTANCE FROM EARTH:  1,340 light-years
FINDING FACTOR:  *
"WOW!" FACTOR:    Binoculars:  **     Small telescopes (3″ to 5″):  ****
Medium telescopes (6″ to 8″):  ****

#### Where to Look

Delta Cephei is easily found adjacent to the southeastern corner of the "house of Cepheus." It creates a conspicuous triangle with the stars Zeta and Epsilon Cephei.

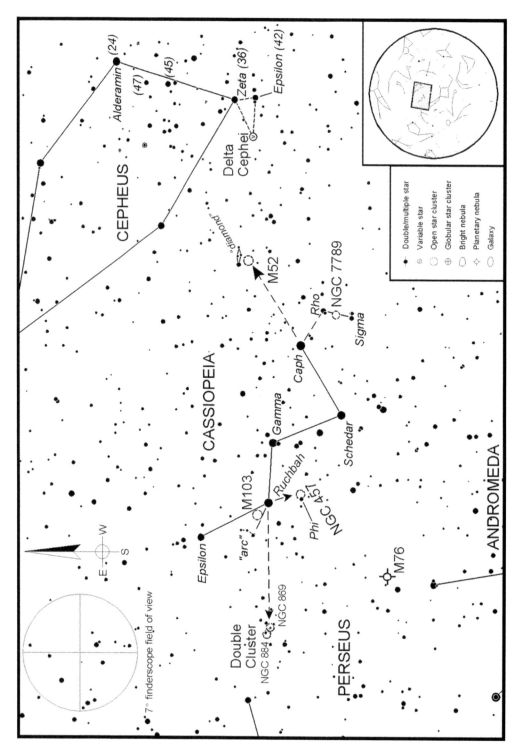

**Figure 8.3** *Many of the objects shown on Autumn Sky Window 1 are circumpolar for much of North America and Europe, letting observers enjoy them year-round.*

### What You'll See

**Through Binoculars.** The subtle golden glint of Delta Cepheus can be readily spotted through binoculars regardless of their size and magnification. Delta is also a variable star, fluctuating in brightness with great regularity.

**Through a Telescope.** Delta displays a gentle yellowish glint through small telescopes. Defocusing the image just a bit should make the color a little more apparent.

Delta attracts attention for a couple of reasons. One is that Delta varies in brightness from magnitude 3.5 to 4.4 over a precisely measured period of 5 days 8 hours 47 minutes. It rises steeply from minimum to maximum brightness in about 1.5 days, while the fall from maximum to minimum brightness is a slow decent across about 4 days.

Take a look at Delta on Autumn Sky Window 1 (Figure 8.3) and notice how some nearby stars have numbers listed in parentheses after them. Each is that respective star's apparent magnitude, expressed in tenths of a magnitude. Alderamin shines at magnitude 2.4, Zeta Cephei at magnitude 3.6, while Epsilon is magnitude 4.2, and so on. (The decimal points have been omitted to avoid being confused for separate stars.) Since none of these stars vary in brightness, they make perfect comparison stars for judging where Delta is in its magnitude cycle.

By comparing a variable star against two or more nearby stars of known, fixed brightness (one brighter and one fainter), some amateurs can actually estimate the variable's unknown magnitude to an accuracy of 0.1 magnitude. For example, let's say that Delta appears brighter than Epsilon but fainter than Zeta. If you judge it to be about halfway between these two other stars in brightness, then your guesstimate for its magnitude will be either 3.9 or 4.0. If it seems closer to Zeta, then your estimate might be more toward magnitude 3.5 or so.

Delta Cephei is a great variable star for beginners because of its quick changes in brightness as well as its visibility. From midnorthern latitudes, Delta stays above the horizon throughout the night year-round. Its range in brightness is also broad enough to be apparent with the slightest optical aid or even the eye alone.

Delta is also a wide double star that is easily resolved through backyard telescopes. The 6th-magnitude companion star lies almost directly south of Delta itself. Astronomers believe that the pair are actually associated with each other and are not just a chance alignment of widely separated stars.

### Delta Cephei

Discovered to be a variable star in 1784 by the British astronomer John Goodricke, Delta Cephei holds a special place in the history of astronomy as the prototype of a special class of yellow-giant variable stars known as Cepheids. All Cepheid variables have very precise correlations between the periods of their magnitude fluctuations and their luminosity, or inherent brightness. By studying this period-luminosity relationship, astronomers can calculate the star's distance very precisely. Cepheids have been observed in star clusters as well as several nearby galaxies, enabling astronomers to calculate the distances to those faraway objects with great accuracy as well.

## NGC 7789: Open Cluster in Cassiopeia

DISTANCE FROM EARTH: 8,000 light-years
FINDING FACTOR:  **
"WOW!" FACTOR:    Binoculars:  *     Small telescopes (3″ to 5″):  **
                 Medium telescopes (6″ to 8″):  ***

### Where to Look

To find NGC 7789, return to the star Caph in the Cassiopeia W, then turn sharply toward the southwest, to Rho Cassiopeiae, which might be too faint to see by eye, but it should be easy to spot through finderscopes and binoculars. From Rho, head south-southeast to Sigma Cassiopeiae, the brightest of three stars in a slender triangle. Along the way, keep an eye out for the faint glow of NGC 7789, about halfway between Rho and Sigma.

### What You'll See

**Through Binoculars.** As with the view through finderscopes, most binoculars will show NGC 7789 as a mist of starlight. The cluster itself is too densely packed and its stars too dim for most binoculars to resolve, causing their combined light to blend together into a soft glow.

**Through a Telescope.** Just as the cluster's density foils resolution through binoculars, so too does it challenge small telescopes. A 3-inch (7.5-cm) telescope will probably fail to show cluster stars, instead displaying a "graininess" across the cluster's misty, circular form. Unlike many star clusters that appear brighter toward the center and dimmer toward the edges, NGC 7789 seems pretty much uniformly bright from edge to edge. A 4-inch (10-cm) should show a few feeble points of light within the cluster, while 6-inch (15-cm) and larger telescopes resolve more and more individual suns.

### NGC 7789

NGC 7789 is famous for being one of the most densely packed open clusters north of the celestial equator. Current estimates put the total population of stars at almost a thousand, all crammed into an area less than 40 light-years across.

At the same time, NGC 7789 is a senior citizen among open star clusters, at about 1.6 billion years old. We know this from studying its individual stars. All stars in a star cluster form at about the same time, although the more massive ones consume the hydrogen fuel in their cores far more rapidly. As this happens, these stars evolve off the Main Sequence and into red giants, an advanced stage in a star's life. By examining the many red giants seen in NGC 7789, astronomers can estimate the cluster's age.

## M103: Open Star Cluster in Cassiopeia

DISTANCE FROM EARTH: 8,500 light-years
FINDING FACTOR:  **
"WOW!" FACTOR:    Binoculars:  *     Small telescopes (3″ to 5″):  **
                 Medium telescopes (6″ to 8″):  **

### Where to Look

M103 can be found just to the northeast of the star Ruchbah in Cassiopeia. Center the star in your finderscope, then look to the northeast for a slender arc of three faint stars. M103 lies in between the arc and Ruchbah and should be bright enough to be at least suspected through a finder.

Another possibility is to place Ruchbah in the southwestern corner of your lowest-power eyepiece, then wait a moment as Earth turns the sky. M103 should enter the northeastern corner of the field within a minute or two.

### What You'll See

**Through Binoculars.** A small, relatively unimpressive object, M103 can be detected through binoculars as a dim glow around a few faint stars.

**Through a Telescope.** M103 has a luster through modern telescopes that Messier could never have imagined through his comparatively poor instruments. Today's amateurs record it as a sparkling collection of stardust set in the pattern of an arrowhead. Marking the tip of the arrowhead is an attractive triple star listed as Struve 131. Look for two 10th-magnitude stars near the system's 7th-magnitude primary sun. Most authorities feel that the association between Struve 131 and the cluster is purely circumstantial, with the star lying between M103 and us. Of the more than one hundred true members in the cluster, the brightest is a 9th-magnitude star lying just southeast of the cluster's center.

## NGC 457: Open Cluster in Cassiopeia

NICKNAMES: Owl Cluster, Dragonfly Cluster, or ET Cluster
DISTANCE FROM EARTH: 9,000 light-years
FINDING FACTOR: *
"WOW!" FACTOR:  Binoculars: **  Small telescopes (3″ to 5″): ***
  Medium telescopes (6″ to 8″): ***

### Where to Look

NGC 457 is a delightful open cluster for deep-sky observers because of its inherent beauty and because it is so easy to find! Begin your hunt at the Cassiopeia W. Trace a line from Epsilon through Ruchbah, and continue it about half again (about 2°) farther toward the southwest. There, through your finderscope or binoculars, you should see a star called Phi Cassiopeiae and, extending to its northwest, the dim glow of NGC 457.

### What You'll See

**Through Binoculars.** Looking like a close-set pair of suns, Phi Cassiopeiae and another, slightly fainter star cataloged as HD 7902 are easy to spot through binoculars. But take a closer look and you will see the faint, combined glow of the other stars in NGC 457 fanning out to their northwest.

**Through a Telescope.** NGC 457 (Figure 8.4) is a fun object to see through just about any telescope. Unlike many other clusters, whose stars are scattered in

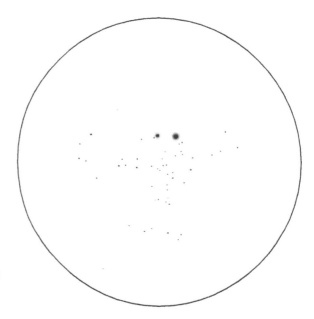

**Figure 8.4** *NGC 457, the Owl Cluster, as drawn through the author's 8-inch reflector at 120×. South is toward the upper right.*

a haphazard manner, those in NGC 457 seem to create a familiar pattern. Some imagine a dragonfly, while others see ET, the Extra-Terrestrial. To many others and to me it looks most like a big-eyed owl, leading to its popular nick-name, the Owl Cluster. The body is drawn from about a dozen faint stars, with another pair marking the tail feathers. Two arcs, each containing about half a dozen suns, form the wings. But what really draws immediate attention are the dazzling "eyes" fashioned from Phi Cassiopeiae and HD 7902.

The Owl Cluster is fairly small, and so handles medium magnifications well. Use a low-power eyepiece to find it at first and to spot the owl pattern, then switch to about 80× for a close-up study.

## NGC 869 and NGC 884: Open Clusters in Perseus

NICKNAME: Double Cluster
DISTANCE FROM EARTH: NGC 869: 7,100 light-years, NGC 884: 7,400 light-years
FINDING FACTOR: *
"WOW!" FACTOR:   Binoculars: ****      Small telescopes (3″ to 5″): ****
                Medium telescopes (6″ to 8″): ***

### Where to Look

NGC 869 and NGC 884 are better known as the "Double Cluster," a striking pair of open clusters. Even from suburban skies, both can be seen with the unaided eye as a faint, elongated smudge of light situated about halfway between the W of Cassiopeia and the "tip" of Perseus. If you can't spot the blur with just your eyes, look through your binoculars or finderscope and

*Owl Cluster*

NGC 457 holds about two hundred stars within its gravitational grip, which spreads across for about 30 light-years.

There remains much controversy around the star Phi Cassiopeiae. Is it a true member of NGC 457 or is it simply a star closer to Earth that happens to fall along the same line of sight? For a star to appear as bright as Phi does from that incredible distance, it must be extremely luminous. In fact, it would have to be about 250,000 times more luminous than our Sun, making it one of the brightest stars in our galaxy. It appears to be moving through space in the same general direction as the cluster, which is usually considered strong evidence for a physical association.

What about the other "eye" star, HD 7902? Although it is a little fainter than Phi, it would still have to be one of the most luminous stars ever found to shine as brightly as it does from that distance. Whether or not either or both of these suns actually belongs to NGC 457 remains a mystery, but regardless, they certainly add to the group's radiance.

extend a line from Gamma Cassiopeiae, the center star of the W, through Ruchbah and continue it toward the east. If you maintain a straight course, you will see both clusters as two tiny knots of stars.

### What You'll See

**Through Binoculars.** Through binoculars, both clusters erupt into vast collections of stars set against a strikingly beautiful field. Most of the stars look either white or blue-white, although you might notice a few orangish suns, as well.

**Through a Telescope.** Few autumn clusters compare to either NGC 869 or NGC 884, but when both are added together, the view literally overflows with stars. Unlike many star clusters, however, where large apertures are used to good advantage, NGC 869 and 884 are best enjoyed through smaller telescopes. Although the clusters are still impressive, larger telescopes tend to dilute the view too much. Regardless of aperture, you'll need your lowest power eyepiece to squeeze both clusters into the same view, since they span about 1°, the same amount of sky as two Full Moons side by side.

Of the two clusters, NGC 869 (the westernmost of the pair, the one closer to Cassiopeia) appears more densely packed. Three- and 4-inch (7.5- and 10-cm) telescopes show between thirty and forty stars clumped together with a few stragglers extending out in all directions. The cluster's fainter stars, though too dim to resolve, blend their light to cast a gentle mist of starshine. You should be able to see some two dozen stars in NGC 884, the eastern cluster in the pair, through small backyard telescopes. Its stars are more loosely packed.

Take a careful look and you will see several yellow and orange stars buried among the many blue-white suns in and around the clusters. One is located very nearly in the center of NGC 884, next to a tight knot of four stars. Two others are found toward the eastern edge of the cluster, while another pair can be found in between NGC 869 and 884. To make subtle star colors stand out more vividly, try defocusing the image just slightly.

### Double Cluster

Whoever discovered the Double Cluster is lost to history, although we do know that their combined presence attracted the eyes of stargazers as long ago as the second century B.C., when Hipparchus mentioned them in his notes. Messier apparently never saw, or at least never recorded, the Double Cluster. How he could miss it but still include M40 (a double star in Ursa Major) and M73 (four stars in Aquarius) is difficult to understand.

Are the two clusters actually physically linked to each other? Studies show that they are near each other in space. NGC 869 (the westernmost cluster) consists of about 200 suns, while NGC 884 holds about 150 stars within its grasp. Both are primarily composed of hot type A and type B supergiant, superluminous stars. Several colorful red supergiant stars are seen in NGC 884 but are conspicuously absent in NGC 869.

## Autumn Sky Window 2

Autumn Sky Window 2 (Figure 8.5) covers the area immediately north and east of the Great Square of Pegasus the Flying Horse. Although this area is relatively empty of naked-eye stars, spotting the two objects showcased here is surprisingly easy thanks to some fairly prominent star patterns.

### M15: Globular Cluster in Pegasus

DISTANCE FROM EARTH: 33,600 light-years
FINDING FACTOR:    *
"WOW!" FACTOR:    Binoculars: **    Small telescopes (3" to 5"): ***
Medium telescopes (6" to 8"): ****

#### Where to Look

From the Great Square, follow the horse's neck and head, marked by the stars Markab, Homan, Biham, and finally, Enif, the star at the tip of Pegasus's nose. M15 is easily found about 4° (half a finderscope field) to the northwest of Enif. Look for a 6th-magnitude star just to the cluster's west, which helps to mark its location. In fact, through finderscopes, the pair almost seem to create a double star, although M15 will look fuzzy even at low power.

#### What You'll See

**Through Binoculars.** Shining at 6th magnitude, M15 is considered to be one of the showpiece objects of the autumn sky. Most binoculars reveal it as a misty patch absent of stars but with a brighter middle.

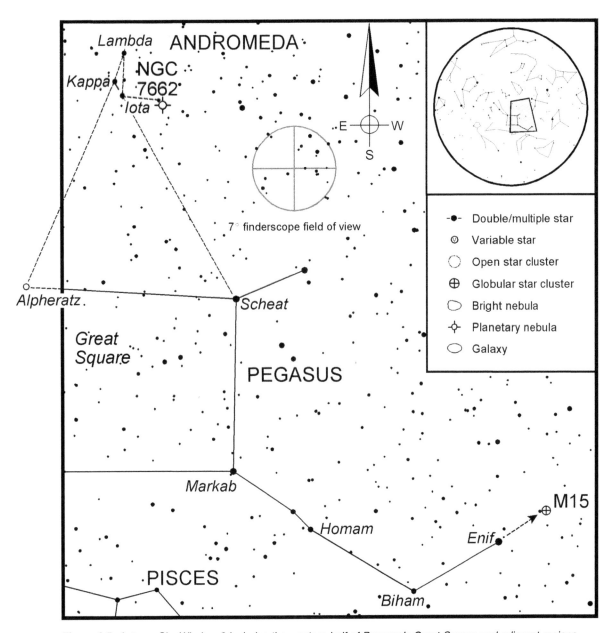

**Figure 8.5** *Autumn Sky Window 2 includes the western half of Pegasus's Great Square and adjacent regions.*

**Through a Telescope.** Even a 2-inch (5-cm) telescope will easily show M15 (Figure 8.6) as a fairly bright, fuzzy ball of starlight. A 4-inch (10-cm) telescope is probably the smallest aperture capable of resolving some of the stars around the edges of this globular. As aperture increases, so will the resolution, with more and more stars seen around and across the densely packed core. Try a moderate-to-high-power eyepiece (say, over 100×) for the best view if seeing conditions permit.

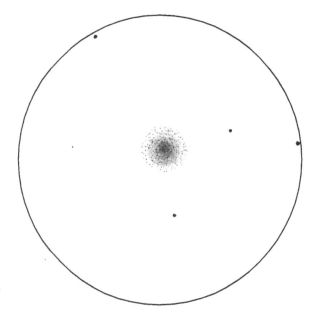

**Figure 8.6** *Globular cluster M15, one of the season's finest, drawn through the author's 8-inch reflector at 120×. South is up.*

### M15

M15 was first spotted in 1746 by Jean-Dominique Maraldi, with Messier rediscovering it in 1764. Its 100,000 or so stars span about 120 light-years.

M15 was the first globular cluster found to contain a planetary nebula. It was first cataloged as Kuster 648 by the German astronomer Friedrich Kuster in 1921, although he erroneously listed it as just another star in the cluster. Seven years later, studies by Francis Pease of Mount Wilson Observatory showed that Kuster 648 was actually a planetary nebula and has subsequently become known as Pease 1.

## NGC 7662: Planetary Nebula in Andromeda

NICKNAME: Blue Snowball
DISTANCE FROM EARTH: 2,200 light-years
FINDING FACTOR: ***
"WOW!" FACTOR:    Binoculars: **    Small telescopes (3″ to 5″): ****
Medium telescopes (6″ to 8″): ****

### Where to Look

Although NGC 7662 lies within the constellation Andromeda, it's easiest to find by using the stars Alpheratz and Scheat along the northern side of Pegasus's Great Square. Trace a line to the northwest of Alpheratz and northeast of Scheat, to create a large triangle with its peak at Kappa Andromedae. Aim at the northern tip of this triangle, where you should find a smaller, skinny triangle of three stars: Lambda, Kappa, and Iota Andromedae. Center your finderscope on Iota, then nudge your telescope about half a finder field due west for NGC 7662.

### What You'll See

**Through Binoculars.** NGC 7662 is a challenging planetary nebula that is bright enough to be seen in 7× glasses, although its 30 arc-second disk will appear stellar. Only its soft blue-green glow sets it apart from surrounding stars. The 13th-magnitude central star of NGC 7662 is well below binoculars' threshold of visibility.

**Through a Telescope.** Even the smallest telescopes show why NGC 7662 is nicknamed the Blue Snowball. Its perfectly circular disk glows with a pale but distinctive bluish cast, growing in intensity as aperture increases. The snowball looks evenly illuminated in most telescopes, although high-power views through 8-inch (20-cm) and larger instruments suggest brighter regions toward the northeast and southwest edges. Spotting NGC 7662's 13th-magnitude central star proves very difficult because the nebula itself is so bright.

### Blue Snowball

NGC 7662 looks like a typical planetary nebula on the surface, with an oval disk surrounding its progenitor star. Photographs taken through the Hubble Space Telescope, however, reveal what are called FLIERs, an acronym that stands for fast low-ionization emission regions. Many believe that these FLIERs are pockets of dense gases that were ejected from the nebula's central star before it created the nebula itself. Not everyone is convinced of that sequence of events, however, with some believing that these features were ejected outward from the central star in the very recent past, perhaps only a thousand years ago.

## Autumn Sky Window 3

Autumn Sky Window 3 (Figure 8.7) includes portions of Aquarius the Water Bearer and Capricornus the Sea-Goat, both members of the sky's "water region." Unfortunately, neither pattern stands out terribly well, especially from light-polluted skies. If you can't find these dim constellations, go back to the autumn star map (Figure 8.1) and notice how the east wing of Aquila the Eagle seems to point southward toward Capricornus. Low-power binoculars will also help spot the faint stars that frame both of these constellations.

### M2: Globular Cluster in Aquarius

DISTANCE FROM EARTH: 37,500 light-years
FINDING FACTOR: **
"WOW!" FACTOR:    Binoculars: **    Small telescopes (3″ to 5″): ***
                 Medium telescopes (6″ to 8″): ***

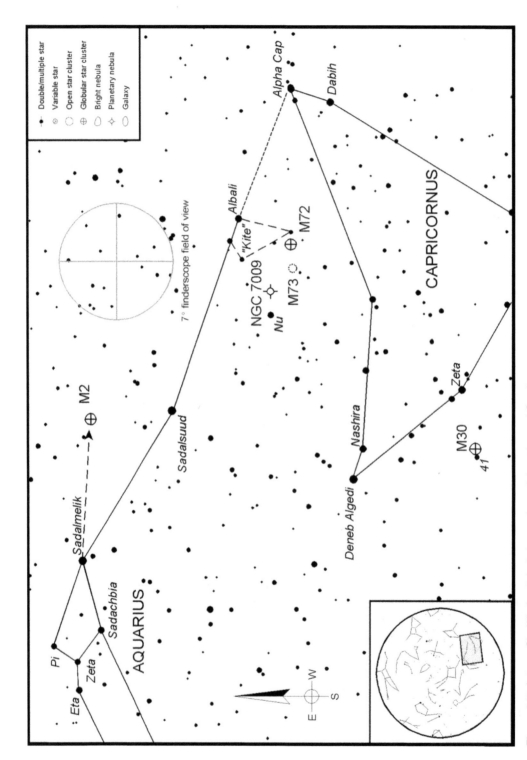

**Figure 8.7** *Autumn Sky Window 3 includes portions of Aquarius and Capricornus.*

### Where to Look

To find M2, first locate the Water Jar pattern of four stars in Aquarius. The Water Jar is made up of the stars Sadachbia and Zeta, Eta, and Pi Aquarii. You will see it on the autumn seasonal star map (Figure 8.1) just to the south of the star Enif, the nose of Pegasus. Flipping back to Autumn Sky Window 3, notice how the Water Jar along with the star Sadalmelik forms a westward pointing arrow? That arrow points toward M2.

### What You'll See

**Through Binoculars.** M2 is visible through nearly all binoculars as a fuzzy "star." Large glasses will show that M2 is not perfectly round but instead is slightly oval.

**Through a Telescope.** Although M2 is one of the brightest globular clusters in the autumn sky, its individual stars are more difficult to resolve than some fainter clusters. Three- and 4-inch (7.5- and 10-cm) telescopes show a few stragglers around the cluster's outer edges, but even an 8-inch (20-cm) can't crack the vault of the cluster's densely packed core. It takes at least a 12-inch (30-cm) to fully resolve stars across the globular's elliptical disk.

## M72: Globular Cluster in Aquarius

DISTANCE FROM EARTH: 53,000 light-years
FINDING FACTOR: ***
"WOW!" FACTOR:     Binoculars: *          Small telescopes (3″ to 5″): **
                   Medium telescopes (6″ to 8″): **

### Where to Look

Start your hunt for M72 by locating the triangular body of the constellation Capricornus and the arrowhead-shaped Water Jar asterism in neighboring Aquarius. Trace from the Water Jar to the star Sadalsuud, then draw an imaginary line to the star Alpha Capricorni, marking the western point of the Capricornus triangle. If you are viewing from the city or suburbs, you may not be able to see these with the naked eye, so as your guide use your finderscope, or better still a pair of binoculars. About halfway along that line from Sadalsuud to Alpha Cap, you should see a long, thin triangle of three stars. The triangle's brightest star, Albali, should be visible without optical aid from most reasonably dark sites. A fourth star to the triangle's south joins to create a kite-shaped pattern of stars. Aim at the bottom star in the kite and look just to the east to find M72.

### What You'll See

**Through Binoculars.** Rated 9th magnitude, M72 is barely detectable as a slightly fuzzy point of very dim light in 7× glasses, but only if used under very dark skies. The slightest bit of light pollution may well mask it completely. Giant binoculars make it a bit more obvious but do little else to improve the visual impression.

**Through a Telescope.** M72 is the faintest of the twenty-nine globular clusters listed in the Messier catalog, and as such, it may prove difficult to catch even through a telescope. Three- and 4-inch (7.5- and 10-cm) telescopes show it as a small, round, nebulous patch of light with a subtle hint of "graininess." These smaller apertures only show the cluster's brighter central core, but 6- and 8-inch (15- and 20-cm) telescopes expand M72's apparent diameter by adding some of its fainter, outer constituents. These same instruments can also resolve a few of its individual stars. Due to its dimness and small appearance, M72 is best viewed at 100× and higher.

## M73: Asterism in Aquarius

DISTANCE FROM EARTH: 2,000 light-years
FINDING FACTOR:  ****
"WOW!" FACTOR:    Binoculars: *        Small telescopes (3″ to 5″): *
                 Medium telescopes (6″ to 8″): *

### Where to Look

M73 is a little less than 1.5° due east of M72. Therefore, the easiest way to find M73 is to find M72 first, then do *nothing*. By letting the stars drift across the field, M73 will come into view in about 6 minutes.

### What You'll See

**Through Binoculars.** The stars that make up M73 can be spotted through steadily held binoculars by using averted vision. Look for M73 as a small patch of hazy starlight just to the east of M72.

**Through a Telescope.** Through a telescope, you can see M73 (Figure 8.8) as a Y-shaped pattern of four stars, ranging in brightness from magnitude 10.3 to 12.3, nestled close together. Through 3- and 4-inch (7.5- and 10-cm) telescopes, their light seems to blend together to give the pattern a misty or nebulous appearance, which is exactly how Messier himself described it in his log. M73's cloudy appearance is quickly dispelled as aperture and magnification increase.

### M73

While viewing M72 in 1780 (discovered by Pierre Mechain that same year), Charles Messier stumbled upon this small group of four faint stars. Whether or not the four stars that constitute M73 are truly, physically related or simply a chance alignment is still under debate among professional astronomers. Some feel that we are seeing the remnant of an open cluster that has all but dissipated over time, while others remain unconvinced. Unfortunately, the otherwise accurate data from the *Hipparchos* astrometric satellite was unable to analyze the stars' distances due to their apparent nearness (as seen from Earth) to one another.

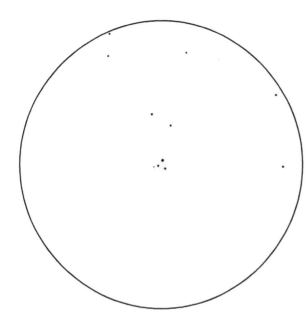

**Figure 8.8** *M73, one of the least impressive Messier objects, is just a pattern of four stars, yet it can give the illusion of nebulosity through smaller telescopes. This drawing was made through the author's 4-inch refractor at 145×. South is up.*

## NGC 7009: Planetary Nebula in Aquarius

NICKNAME: Saturn Nebula
DISTANCE FROM EARTH: 2,400 light-years
FINDING FACTOR: ***
"WOW!" FACTOR:    Binoculars: *      Small telescopes (3″ to 5″): **
                 Medium telescopes (6″ to 8″): ***

### Where to Look

NGC 7009 is most readily found by first locating the "kite" asterism that was used to find M72 and M73, then looking for Nu Aquarii, a lone star about half a finderscope field to the east of the kite. Nu just might be visible to the naked eye from a reasonably dark location. NGC 7009 is 1° to the west.

### What You'll See

**Through Binoculars.** NGC 7009 is visible in binoculars as a greenish point of light thanks to its unusually high-surface brightness.

**Through a Telescope.** The appearance of NGC 7009 (Figure 8.9) through a small, low-power telescope is similar to that seen through binoculars; that is, a small, pale blue-green, starlike point. Only by increasing magnification beyond about 50× will the nebula's tiny disk shape come into view. A 4-inch (10-cm) telescope operating at 100× shows NGC 7009 to be almost football shaped. That same telescope can also show the nebula's 11th-magnitude central star, which is quite bright as central stars go.

If you get a chance, take a look at NGC 7009 through an 8-inch (20-cm) or larger telescope. These add a pair of peculiar extensions, sometimes called

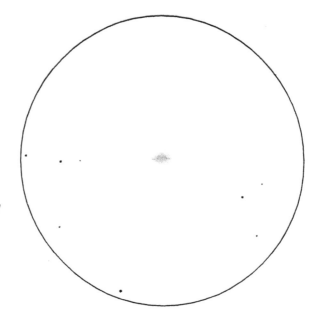

**Figure 8.9** *The Saturn Nebula, NGC 7009, is bright enough to be seen through small backyard telescopes, though its ringlike extensions will take at least a 4-inch telescope to be seen. This drawing, oriented with south up, was made through the author's 8-inch reflector at 200×.*

"antennae," extending away from the nebula's disk along its long (or major) axis. It was this peculiar appearance, which vaguely resembles the planet Saturn with its ring system, that gave rise to NGC 7009's popular nickname, the Saturn Nebula.

### Saturn Nebula

NGC 7009 was discovered by William Herschel on September 7, 1782. It's actually rather surprising that Messier, Mechain, and their contemporaries missed seeing it first, since NGC 7009 stands out so well. This is due in part to its rather barren surroundings, but mostly thanks to its remarkable surface brightness. Its high luminosity is caused by unusually strong ultraviolet radiation emitted by its central star. The nebula's bluish green color is created by doubly ionized oxygen, a product of the process.

Studies with the Hubble Space Telescope suggest that the nebula's central star first expelled an oval disk of material, which now acts to direct stellar winds into two opposing jets, creating the two antennae. Unlike many planetary nebulae, which have a complex, swirling structure, NGC 7009 appears to be strangely free of internal turbulence.

## M30: Globular Clusters in Capricornus

DISTANCE FROM EARTH: 26,100 light-years
FINDING FACTOR: ***
"WOW!" FACTOR:    Binoculars: *    Small telescopes (3″ to 5″): **
                 Medium telescopes (6″ to 8″): **

### Where to Look

M30 is situated in the southeastern part of the constellation Capricornus in a rather barren section of the early autumn sky. Getting there can be a challenge. The best way is to trace out the triangular shape of Capricornus first. Since the constellation rides so low in the southern sky from midnorthern latitudes, you will need a dark view that is free of obstructions and other interferences in that direction. The stars Nashira and Deneb Algedi mark the Sea-Goat's tail and the eastern point of the Capricornus triangle. Turn southwest toward the triangle's southern tip, pausing partway at the 4th-magnitude star Zeta Capricorni. Center Zeta in your finderscope, then look toward the eastern edge of the field. A 5th-magnitude star, 41 Capricorni, should just be coming into view. Center on 41, then look through your telescope. M30 is just to that star's west, easily fitting into the same eyepiece field.

### What You'll See

**Through Binoculars.** M30 shines at 7th magnitude. Given a good view, it should be visible as a dim, round patch of grayish light drawing to a brighter core.

**Through a Telescope.** M30 (Figure 8.10) is one of the many fine Messier objects that is often forgotten because of its remote location in the sky. Those who take the extra effort to find it, however, are treated to a pleasant object, even

**Figure 8.10** *M30 in Capricornus may have undergone a collapse in the distant past, creating an exceptionally dense core. South is up in this photograph by George Viscome through a 14.5-inch f/6 reflector.*

through small apertures. A 4-inch (10-cm) telescope reveals a bright, fuzzy core engulfed in a fainter outer cloud of unresolved stars, while a 6-inch (15-cm) begins the job of stellar resolution by showing hints of dim member stars buried within the outer haze.

## Autumn Sky Window 4

Autumn Sky Window 4 (Figure 8.11) includes most of the constellation Andromeda as well as tiny Triangulum the Triangle and a portion of western Perseus. Andromeda should be fairly easy to spot, in part because of its position near the zenith at this time of year and in part because the prominent W pattern of Cassiopeia lies just above (that is, to its north). Trace out the pattern of Andromeda as it extends off the Great Square of Pegasus, then confirm which stars are which by comparing their positions relative to Cassiopeia.

### M31: Spiral Galaxy in Andromeda
### M32: Elliptical Galaxy in Andromeda
### M110: Elliptical Galaxy in Andromeda

NICKNAME: Andromeda Galaxy
DISTANCE FROM EARTH: 2.9 million light-years
FINDING FACTOR:  *
"WOW!" FACTOR:   Binoculars: ****      Small telescopes (3″ to 5″): ****
                 Medium telescopes (6″ to 8″): ****

#### Where to Look

M31, the famous Andromeda Galaxy, is easily found once you locate its home constellation, which extends to the northeast of the Great Square of Pegasus. Start at Alpheratz, the star that appears to be shared by both the Great Square and Andromeda, then find the group of stars that arcs from Alpheratz through Delta Andromedae and Mirach to Almach. Center your finderscope on Mirach, then take a turn to the northwest, to the faint naked-eye star Mu Andromedae. Continue farther north to Nu Andromedae, which is fainter still but readily visible through finders and binoculars. Look to the west of Nu for a large smudge of light. That's M31. In fact, from even moderately dark skies, M31 is visible to the naked eye, holding the distinction of being the most distant sight visible without telescopic aid.

#### What You'll See

**Through Binoculars.** Binoculars show the Andromeda Galaxy as a broad, oval smudge of grayish light that, at first, may appear rather dull and featureless. Closer inspection, however, reveals a pronounced core highlighting the center of the disk. At the same time, the full extent of M31's oval glow is oriented with the long dimension (or *major axis*) running northeast-southwest for an incredible 3° to 5°—probably more than half the field of view of your binoculars! That's a distance equal to ten Full Moons stacked end to end! No other

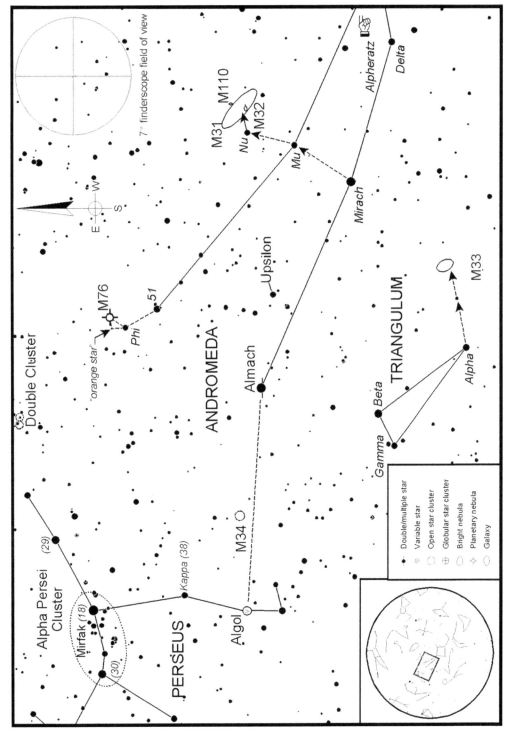

**Figure 8.11** *Autumn Sky Window 4 holds several favorite autumn objects, including the major galaxies closest to our own.*

galaxy visible from the Northern Hemisphere can compare, though to see its full extent, you will need dark skies and steadily supported binoculars.

Two of M31's satellite companion galaxies can also be glimpsed through binoculars, though both will test your skills. The smaller and brighter of the pair, M32, can be spotted as a small, almost starlike patch of light due south of M31's central core, while the other companion, M110, is larger and fainter, and so is more difficult to spy. Look for it to the north of M31's core, positioned about twice the distance away as M32.

**Through a Telescope.** Through most backyard telescopes, M31 looks like a bright, elongated glow that extends outward toward the edge of the eyepiece field, as shown in Figure 8.12. Low power is an absolute must for the best view, since any extra magnification will only limit the field size even more. Regardless of your telescope, a 25- to 40-mm eyepiece will probably be your best choice, although with a long focal-length telescope like an 8-inch (20-cm) f/ 10 Schmidt-Cassegrain, even these will prove very restrictive.

Look a little to the south of M31's bright core and you might see a second, smaller, fainter oval patch of light. That is M32, one of M31's small companion galaxies. Look for it at the right angle of a triangle created with two stars in the Milky Way. Next, move north, placing M31's core just on the edge of the field, and look for the larger, fainter glow of M110, a second companion galaxy. Be warned: M110 is very dim, making it a tough catch in 3- and 4-inch (7.5- and 10-cm) telescopes, especially through light pollution. It looks about twice as large as M32 but is missing any central concentration.

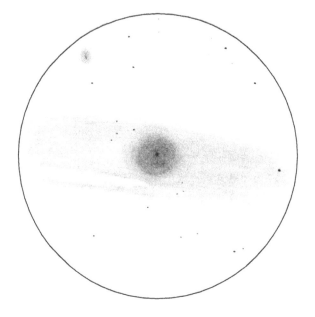

**Figure 8.12** *M31, the Andromeda Galaxy, as perceived through the author's 8-inch reflector at 55×. The satellite galaxy M32 is seen toward the upper left in this south-up view. Compare this drawing to the photograph in Figure 8.13.*

## *Andromeda Galaxy*

The Andromeda Galaxy was first recorded in the tenth century when it was noted by the Persian astronomer Al-Sufi. The German astronomer Simon Marius became the first person to view M31 telescopically back in 1612, although he certainly did not recognize its true nature as a separate galaxy. Nearby M32 was first spotted in 1749 by Guillaume Le Gentil, while M110 was found by Charles Messier in 1773. Curiously, both M32 and M110 were included on a drawing of M31 made by Messier, yet he never included M110 in his listing published in 1807. No one quite knows why. M110 was finally added to the "official" Messier catalog 159 years later.

M31 (Figure 8.13) is classified as an Sb spiral, indicating moderately tightly wound spiral arms. Believed to be 200,000 light-years across, the Andromeda Galaxy proves to be twice as large as our own Milky Way, but it is less densely packed. The Hubble Space Telescope has also revealed that it appears to have an unusual double nucleus. Why remains unclear, although it might only be an effect caused by opaque dust obscuring a portion of a single core, resulting in an optical illusion.

Both M32 and M110 are excellent examples of dwarf elliptical galaxies. M32 appears almost perfectly round and is classified as an E2 elliptical. It is thought to measure 8,000 light-years across, a very small fraction of M31. M110 is an E6 elliptical galaxy because of its more oblate profile. M32 and M110 are the two brightest of as many as ten companion galaxies associated with M31.

**Figure 8.13** *M31, the Andromeda Galaxy, as photographed by George Viscome through a 14.5-inch f/6 reflector. Compare this view to Figure 8.12.*

Okay, time for a little honesty. When they first see M31 through a telescope, many new stargazers come away from the eyepiece a little disappointed. Part of that is probably the result of all the magnificent photographs that show M31 littered with complex clouds and intricate structure within its spiral arms. But when the galaxy is observed through a telescope, none of that is there. What's going on here?

A few things are working against us when it comes to viewing the Andromeda Galaxy. One is that it's too big! As mentioned above, the entire breadth of the spiral arms embraces some 5° of sky. That is *much* larger than most telescopes can handle in a single view. Instead all we can see is a portion of the galaxy, and even that fills the field, which means that image contrast against the background sky is quite low, which is a second problem.

But before you go away disenchanted, stop and think for a moment about what you are seeing. The Andromeda Galaxy is the closest major galaxy to our own, yet it still lies an estimated 2.9 million light-years away. Of course, not only is that an unimaginable distance, but it is also a look back in time. Remember, we are seeing Andromeda not as it looks tonight, but as it did nearly three million years ago! The light we see coming from Andromeda tonight left there when our earliest ancestors began to evolve in Africa. Both that ray of light from M31 and our species have come a long way since!

## M33: Spiral Galaxy in Triangulum

DISTANCE FROM EARTH: 3 million light-years
FINDING FACTOR: ***
"WOW!" FACTOR:    Binoculars: **        Small telescopes (3″ to 5″): **
                 Medium telescopes (6″ to 8″): ***

### Where to Look

Begin at the small constellation Triangulum, a small isosceles triangle of stars tucked in between Andromeda, Aries, and the Pleiades in Taurus. Aim toward the star Alpha Trianguli at the triangle's apex. Viewing through your finderscope, swing your telescope half a finder field toward Alpheratz in Andromeda. You should see a faint field star somewhere near the center, with Alpha Tri now in the eastern half of the field. Move another half a finder field toward Alpheratz, then switch your view to the telescope and take a careful look for a large, very faint glow. That will be M33. Try sweeping back and forth *slightly* if you don't find it at first, rechecking your finder to make sure you don't wander off too far.

Be forewarned that most beginners discover that this can be one tough object to find despite its rather bright magnitude rating of 6.3. The difference in visibility between M31, at magnitude 3.5, and M33 is striking. M31 is *easy* to find, so it stands to reason that M33 should be as well. After all, both are nearby as galaxies go, and both have high magnitude ratings. So, how hard could M33 be?

The answer is plenty hard! The problem is that an object's magnitude value is based on how bright it would appear if it were collapsed to a point of light, like a star. Of course, M33 is not a point of light at all, it's a disk, and a large one at that. As a result, M33 has a very low *surface brightness;* that is, its brightness per unit of area is so low compared to the background sky that the dim glow of the galaxy blends seamlessly into the background. Many people searching for M33 pass right over it without even noticing it, so take your time and scan back and forth slowly, using your lowest-power eyepiece.

### *One of the Big Three*

M33 (Figure 8.14) is one of the Big Three in our local group of galaxies (the others being M31 and the Milky Way). But while M31 is perhaps twice as massive as our galaxy, M33 is somewhat smaller. Classified as a broad-armed Sc spiral galaxy, M33 is tilted almost face-on from our perspective, one of the reasons why its surface brightness is so low. Its spiral arms can be suspected faintly through large amateur telescopes, pinwheeling away from the galaxy's central hub. Several knots of stars and interstellar matter dot the arms. The brightest, NGC 604, was mentioned above as visible through small telescopes, possibly even through binoculars on exceptional nights.

Based on the most recent observations conducted by the *Hipparchos* satellite of the Cepheid variables found within M33, the galaxy has been pushed back to almost exactly 3 million light-years away. This is some 700,000 light-years farther than the previous estimate. Given this new distance, M33 probably measures about 50,000 light-years in diameter, only about half that of the Milky Way.

Most historians credit the discovery of M33 to Giovanni Batista Hodierna, who spotted it some time before 1654. Messier independently rediscovered it in 1764.

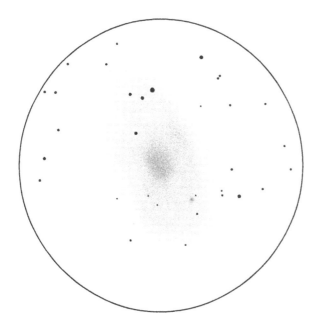

**Figure 8.14** *M33 in Triangulum is the bane for many a new stargazer because of its low surface brightness. This drawing, made through the author's 8-inch reflector at 44×, is oriented with south at the top.*

### What You'll See

**Through Binoculars.** M33 is often easier to find in steadily held binoculars than through telescopes, since their wider fields of view are better at segregating the large galactic disk from the surroundings. Their wide fields also make it possible to fit Alpha Trianguli, M33, and that intermediate 6th-magnitude star all into the same view. Once spotted, M33 looks like an amorphous blur that fades away from a brighter center.

**Through a Telescope.** Again, even though M33 is the second closest major galaxy to our own, it is far more difficult to find than others that are many times farther away. Therefore, we must be smart when looking for it, especially that first time. You should use averted vision through your lowest-power eyepiece for the best view. Even then, M33 will probably look like a ghost at first. A very dim glow will light the eyepiece field, drawing to a brighter central core.

Once you land M33 itself, try your luck with finding an object that is actually inside the galaxy. Telescopes as small as 3 inches (7.5 cm) in aperture are large enough to reveal a second, small glowing patch of light just to the northeast of M33's core. Messier never saw this second object; instead, we know it by its entry number in the *New General Catalog*, NGC 604. In reality, NGC 604 is a massive emission nebula, much larger than the famous Orion Nebula in our winter sky.

## M76: Planetary Nebula in Perseus

NICKNAME: Little Dumbbell
DISTANCE FROM EARTH: 3,400 light-years
FINDING FACTOR: ***
"WOW!" FACTOR:    Binoculars: not resolvable    Small telescopes (3" to 5"): **
                 Medium telescopes (6" to 8"): ***

### Where to Look

Although M76 is listed as one of the faintest Messier objects, it is surprisingly easy to find once you can picture the constellation Andromeda extending away from the Great Square of Pegasus. Follow the line of faint stars that moves northeastward from Alpheratz, the star marking the northeast corner of the Square, and continues on through Mu Andromedae (which we just visited on our way to M31), to end at 51 Andromedae and Phi Persei, a pair of dim naked-eye stars. Aim toward Phi using your finderscope, then look through your telescope using a medium-power eyepiece (around 70× would be good for starters). You should see Phi and a dim, orange star immediately to its north. M76 completes a right triangle with these two stars, the orange sun marking the right angle itself.

### What You'll See

**Through Binoculars.** Unfortunately, M76 is too faint and too small to be readily distinguishable with most binoculars.

**Through a Telescope.** Telescopes with apertures as small as 3 inches (7.5 cm) will show M76 as a tiny, faint point of fuzzy light, which at first glance may look like an ordinary star. But you will need at least 100× before you notice that this so-called star "just doesn't look right." Using 4- and 6-inch (10- and 15-cm) telescopes, many observers describe it as rectangular and elongated approximately north-south. Others say it looks just like a celestial peanut! But most agree that it bears at least a passing resemblance to M27, the famous Dumbbell Nebula in the summer constellation Vulpecula, and so M76 has been nicknamed the Little Dumbbell (Figure 1.3).

It's interesting to point out an important lesson here for all observers, no matter what kind of telescope is being used. Remember M33 just a moment ago? It's listed as magnitude 6.3, which sounds as though it should be fairly bright and easy to find. It is neither. Now, contrast that with M76, listed at magnitude 11.4. Should be faint, right? But M76 is actually easier to find than M33! Why? The answer is surface brightness. M76 has a brighter surface than M33, and is therefore easier to see. The point of all this is *never* to be misled by an object's listed magnitude. In some cases, the listed magnitude makes an object sound much brighter than it actually appears, while in others, it misleads observers into thinking that an object is fainter.

### *Little Dumbbell*

Discovered by Pierre Mechain in 1780, M76 is a classic example of what astronomers call a bipolar planetary nebula. In these nebulae, it is believed that a disk of obscuring dust not only hides the central star from view, but it also channels material into two exhaust plumes rather than streaming away evenly. The result is a planetary nebula that appears to blossom much as a butterfly extends its two wings, which has led some to call them "butterfly" planetary nebulae. This two-part appearance also led to M76 being assigned two entries, NGC 650 and NGC 651, in the *New General Catalog*.

The distance to M76, like that of many planetary nebulae, is not well established. Estimates place it about 3,400 light-years away, although published values range anywhere between 1,700 and 15,000 light-years. If we accept the distance of 3,400 light-years, then M76 spans about 4.5 light-years.

## Almach: Double Star in Andromeda

DISTANCE FROM EARTH: 355 light-years
FINDING FACTOR:  *
"WOW!" FACTOR:    Binoculars: not resolvable        Small telescopes (3″ to 5″): ****
                 Medium telescopes (6″ to 8″): ****

### Where to Look

Almach, also known as Gamma Andromedae, lies at the northeastern end of the line of stars in Andromeda that starts at Alpheratz and passes through Mirach. Shining at 2nd magnitude, Almach is bright enough to be seen with the naked eye even from most light-polluted areas.

### What You'll See

**Through Binoculars.** Unfortunately, even though Almach is visible with the naked eye, binoculars just don't have the "oomph" to show that it is a binary star.

**Through a Telescope.** One of autumn's premier binary stars, Almach is a wonderfully colorful binary star that is easily resolvable through just about any telescope. The brighter of the two stars is an orange giant primary star, while its fainter, hotter companion shines a deep blue. The pair is separated by about 10 arc-seconds, with the bluish B star lying to the northeast of Gamma A.

### *Almach*

The two stars in this system are separated by about 100 billion miles (162 billion km) or about 1,100 astronomical units and take over 1,000 years to complete one orbit. In 1842, Otto Struve discovered that the blue star is itself a close binary star. Its 6th-magnitude companion orbits much more quickly, taking only 61 years to complete a circuit. Unfortunately, right now the two stars are too close to be easily resolved through most backyard telescopes.

## Alpha Persei Cluster:  Star Cluster in Perseus

DISTANCE FROM EARTH: 601 light-years
FINDING FACTOR:  *
"WOW!" FACTOR:    Binoculars: ****    Small telescopes (3″ to 5″): **
                 Medium telescopes (6″ to 8″): *

### Where to Look

To spot the Alpha Persei Cluster of stars, first find Alpha Persei, also known as Mirfak. Draw a line from the Great Square to the bright star Capella in Auriga, just rising in the northeast. (Although neither is shown on Autumn Sky Window 4, both are labeled on Figure 8.1, the seasonal chart at the beginning of this chapter.) About two-thirds of the way along that line, you should see another star, fainter than Capella. That's Mirfak.

### What You'll See

**Through Binoculars.** The Alpha Persei Cluster is one of the sky's true gems for binoculars! This magnificent region is bound to overwhelm you with its beauty no matter what size binoculars you are using. Many dazzling blue-white beacons are scattered throughout the area in small clumps and patterns, with 3rd-magnitude Mirfak being the brightest of the bunch. While most of the stars in the group appear white or blue-white, one or two might show the slightest tinges of yellow or orange.

**Through a Telescope.** As striking as the cluster is through binoculars, most telescopes just cannot do it justice. The problem is the cluster's size, equivalent to three Full Moons stacked end to end. That makes isolating the cluster from its surroundings difficult to do, since most telescopes can squeeze no more than

1°, or two Full Moons, into a single field of view. The exception to this may be short-focal-length refractors and reflectors, which are almost more like one-eyed binoculars than conventional telescopes. But even those will need very low-power eyepieces for the best view.

### *Alpha Persei Cluster*

The Alpha Persei Cluster, discovered before 1654 by Giovanni Batista Hodierna, is one of the closer star clusters to Earth, which explains why it appears so large in our sky. To be technically correct, this grouping is referred to as an OB Association, not an open star cluster. An OB Association contains mostly spectral-type O and B blue-white stars, more loosely gathered than in typical open clusters. In the case of the Alpha Persei group, some fifty stars scattered across 32 light-years are bound by their mutual, though weak, gravitational field. Studies indicate that the cluster's stars are only about 51 million years old, younger than many diamonds found here on Earth. Eventually, OB Associations disperse as the gravitational bonds continue to weaken. (Open clusters disperse as well, but much more slowly.)

## Algol: Variable Star in Perseus

DISTANCE FROM EARTH: 93 light-years
FINDING FACTOR: *
"WOW!" FACTOR:  Binoculars: ****     Small telescopes (3″ to 5″): **
Medium telescopes (6″ to 8″): **

### Where to Look

To find Algol, go back to Mirfak, the brightest star in the constellation Perseus. Recall that Mirfak lies along an imaginary line connecting the Great Square of Pegasus and the bright star Capella, just rising in the northeast. From here, hop southward from Mirfak to fainter Kappa Persei, then finally to Algol.

### What You'll See

**Through Binoculars.** Algol is easily visible in binoculars throughout its entire magnitude cycle. That and its short period make it an ideal variable star to watch through binoculars, telescopes, and even with just the naked eye. With enviable regularity, Algol takes 2 days 20 hours 48 minutes 56 seconds to go from magnitude 2.1 to 3.4, and back again.

Want to try your luck at watching and charting the changing brightness of Algol? Take a look at Autumn Sky Window 4 (Figure 8.11). The magnitudes of several suitable comparison stars are shown next to the stars themselves. (The decimal points have been omitted to avoid confusing them for separate stars.) Make your estimate of Algol's magnitude, then plot it on Figure 8.15, which is a light curve of Algol. A light curve is simply a graph that compares a variable star's magnitude to time. Magnitude is plotted along the vertical axis, while time is plotted horizontally.

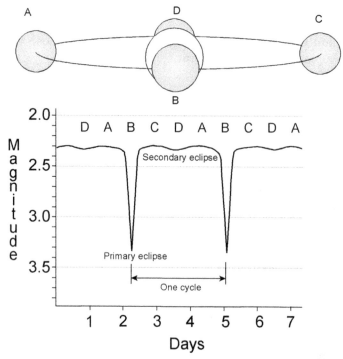

**Figure 8.15** *The "demon star," Algol, varies in brightness on a regular basis as an unseen companion star passes in front of and behind the system's primary sun. The points along the light curve lettered A through D correspond to the stages in the companion's orbit of the primary star.*

### Algol

Early stargazers watching in fear and wonder as Algol changed in brightness immediately assumed that it must be possessed by evil powers. Even it's name, which is Arabic for the "Demon's head," paints an evil picture. Algol is often portrayed in the constellation Perseus as marking Medusa's winking eye. Homer referred to Algol in the *Iliad* as "the Gorgon's head, a ghastly sight, deformed and dreadful, and a sight of woe," while ancient Chinese astronomers called it Tseih She, which translates as "Piled-up Corpses." So while Algol's fluctuating appearance was officially first detected in 1669, these names would certainly seem to indicate that many early cultures knew of its peculiar nature as well.

Today, Algol is one of the sky's most watched variable stars. Algol is an eclipsing binary star system in which a spectral-type B blue-white primary star is orbited by a type K orange companion star. As seen from Earth, the system's orbital plane is tilted almost edge-on, causing the secondary star to pass alternately in front of and behind the primary sun. Although the stars are too close to each other to be resolved separately, we see these passages as fluctuations in the system's total brightness.

The greatest dip in magnitude occurs when the fainter secondary sun passes in front of the primary, temporarily blocking part of its light. An eclipse lasts about ten hours, with Algol at minimum brightness for about two hours. The star then slowly returns to its maximum, preeclipse brightness as the companion continues in its orbit. A secondary dimming amounting to only about 0.1 magnitude occurs as the companion star is eclipsed by the primary.

**Through a Telescope.** Telescopes will not show Algol in any greater detail than binoculars and in fact will make it more difficult to check the variable's brightness against suitable comparison stars because of their more restrictive fields of view. If you want to study Algol but do not own a pair of binoculars, your best bet might be to view it through your telescope's finder or even your eye alone.

## M34: Open Cluster in Perseus

DISTANCE FROM EARTH: 1,400 light-years
FINDING FACTOR: **
"WOW!" FACTOR:    Binoculars: ***    Small telescopes (3″ to 5″): ***
                 Medium telescopes (6″ to 8″): ***

### Where to Look

First, trace out the form of Andromeda, beginning at Alpheratz at the Great Square, then continuing through Mirach to Almach. East of Almach, look for the curved body of Perseus. Center your aim about halfway along a line connecting Almach in Andromeda to Algol. M34 lies just to the north of that line, a little east of the halfway point. M34 is bright enough to be seen through binoculars and finderscopes as a misty patch of light, and might even be visible to the naked eye under very dark, clear skies.

### What You'll See

**Through Binoculars.** An outstanding open cluster, M34 appears as large as the Full Moon in our sky and holds some sixty stars within its gravitational grip. Binoculars show the brightest as twinkling points bathed in the hazy glow of fainter unresolved suns.

**Through a Telescope.** M34 (Figure 8.16) is a wonderful cluster to view through small telescopes. More than two dozen fairly bright stars set against the nebulous backdrop of fainter, unresolved suns pepper the cluster. Many of these stars are set in pairs and clumps. The group as a whole looks decidedly rectangular, with strings of stars threading across the object. Low magnification will give the best view.

# Autumn Sky Window 5

Autumn Sky Window 5 (Figure 8.17) covers Aries the Ram, eastern Pisces the Fishes, and northern Cetus the Whale. None of these constellations stand out from the crowd, and indeed they may be swallowed up by light pollution in urban settings. To orient yourself, find the Great Square of Pegasus and the Pleiades (or Seven Sisters) rising in the east. The three stars that make up the body of Aries lie in between. Once you spot Aries, use its stars as pointers to find the other patterns in this area.

**Figure 8.16** *Open star cluster M34 in Perseus. South is up in this photograph taken through a 14.5-inch f/6 reflector by George Viscome.*

## Mesarthim: Binary Star in Aries

DISTANCE FROM EARTH: 204 light-years
FINDING FACTOR:  *
"WOW!" FACTOR:    Binoculars: not resolvable     Small telescopes (3″ to 5″): ****
Medium telescopes (6″ to 8″): ****

### Where to Look

Mesarthim, also known as Gamma Arietis, is visible to the naked eye as the faintest and southernmost of the three main stars in Aries.

### What You'll See

**Through Binoculars.** Although easily visible through binoculars, Mesarthim masks its double personality at the low magnifications of most binoculars.

**Through a Telescope.** By increasing magnification to about 50×, even a 2-inch (5-cm) telescope will show Mesarthim as two pearly white 5th-magnitude suns. Separated by just under 8 arc-seconds, the stars are positioned exactly north-south of each other.

## M74: Spiral Galaxy in Pisces

DISTANCE FROM EARTH: 35 million light-years
FINDING FACTOR: ****
"WOW!" FACTOR:    Binoculars: not resolvable     Small telescopes (3″ to 5″): *
Medium telescopes (6″ to 8″): **

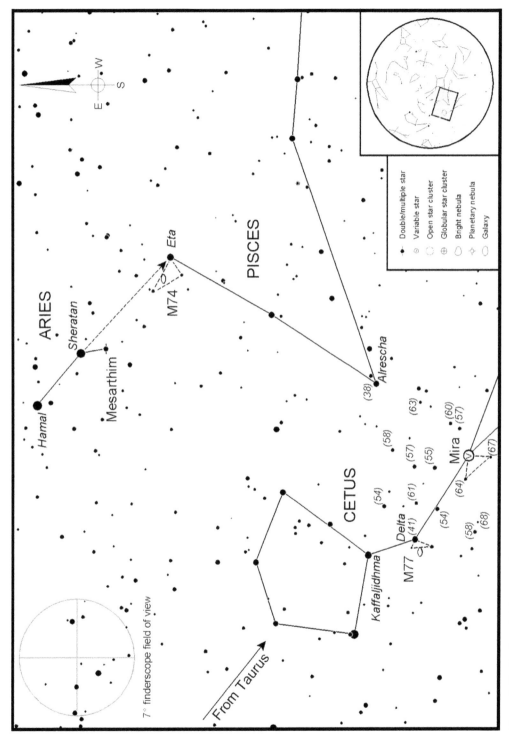

**Figure 8.17** *Far from bright stars, Autumn Sky Window 5 includes two of the most difficult Messier objects to find.*

### Where to Look

Okay, this is a tough one! In fact, it is my humble opinion that M74 is the most difficult object in the entire Messier catalog. (How Messier could have included this object in his catalog and not the Double Cluster in Perseus is beyond me, but I digress.) To spot M74, first locate the small constellation Aries the Ram, set east of the Great Square and just west of the Pleiades in Taurus the Bull. Draw a line from Hamal to Sheratan, the two brightest stars in Aries, and extend it to the southwest. Viewing through your finderscope, follow the line to the star Eta Piscium in Pisces the Fishes. You might also be able to see Eta with the naked eye, though rather faintly. From Eta, look for a small triangle created with two fainter stars that seems to point back toward Aries. A "double star" marks the tip of the triangle, offering a hint that you are in the right place. M74 lies inside this triangle. Look for its vague glow with your lowest-power eyepiece.

### What You'll See

**Through Binoculars.** M74 is too faint to be seen through binoculars under most circumstances.

**Through a Telescope.** Are you up to the M74 challenge? If you are a new star-gazer, my advice is to wait until you get a few easier objects under your belt before trying to spot this difficult target. And even then, remember the two *P*'s: patience and perseverance.

A face-on spiral galaxy, M74 is noted for its nearly stellar central core and very faint spiral arms. Sadly, M74 will never look as nice visually through a telescope as is does in photographs. Instead, expect to see only a dim glow surrounding a slightly brighter core through medium-size telescopes. Even a trace of light pollution may obliterate any indication of the spiral halo, leaving only a starlike point.

Once you know that you are aimed at the right field, switch to medium power to isolate the galaxy from its surroundings. Always use averted vision when looking for extremely faint targets like M74. If that doesn't bring it out, try jiggling or lightly tapping the side of your telescope. The eye can detect very low surface-brightness objects more readily if they are moving back and forth slightly instead of just sitting motionless. Once you spot M74, be sure to steady the image for a better view.

## M77: Spiral Galaxy in Cetus

DISTANCE FROM EARTH: 60 million light-years
FINDING FACTOR:  ***
"WOW!" FACTOR:   Binoculars: not resolvable    Small telescopes (3″ to 5″):  **
                Medium telescopes (6″ to 8″):  **

### Where to Look

Cetus is a large, vague constellation riding along the southern portions of the autumn evening sky. Spotting it can prove a real challenge, especially from suburban and urban observing sites. My best suggestion is to wait until the winter constellation Taurus is up in the sky. The V-shaped head of Taurus, discussed in the next chapter, points more or less toward the pentagonal pattern of five stars that marks the northern end of Cetus. Although faint, the pattern is distinctive enough to be visible in the suburbs. Aim your telescope its way, then look just to the south of the pentagon for the 4th-magnitude star Delta Ceti. M77 lies just to the east of a small triangle formed by Delta and two fainter field stars.

### What You'll See

**Through Binoculars.** Although it's theoretically possible to see M77 through 70-mm and larger binoculars, this galaxy is really a "telescope-only" object under most circumstances.

**Through a Telescope.** M77 appears through amateur telescopes as a small, oval glow highlighted by a brighter, fairly large central core. A 10th-magnitude star lies just to its east. Of course, this star is inside our Milky Way galaxy and has no physical association with the galaxy.

### M77

Indexed as an Sb spiral galaxy, M77 is a famous "active" galaxy, called a Seyfert galaxy. Seyfert galaxies, named for their discoverer, the American astronomer Carl Seyfert, have starlike nuclei and intense, variable radio noise emissions. By studying its spectrum, astronomers have also concluded that giant clouds of gas are erupting from M77's core, moving at several hundred miles per second. The mechanism that is powering the active core of M77 remains a mystery.

## Mira: Variable Star in Cetus

DISTANCE FROM EARTH: 400 light-years

FINDING FACTOR: **

"WOW!" FACTOR:    Binoculars: **        Small telescopes (3″ to 5″): **
                  Medium telescopes (6″ to 8″): **

### Where to Look

Depending on where it is in its magnitude cycle, Mira may be visible to the naked eye or it may be below the threshold of finderscopes and binoculars. In either case, begin your quest by following the directions for M77 above. Once you spot Delta Ceti, look about one finder field to the southwest. There you *may* see a reddish star that marks the right angle in a triangle formed with two other stars to its east and south. Center your search at that point, then switch to your telescope to confirm Mira. Take care not to be fooled by a 9th-magnitude star that lies just to the east of Mira (more on that star later).

### What You'll See

**Through Binoculars.** Mira is a great variable star to study through binoculars. Over the course of its magnitude cycle, Mira goes from a high of about 3rd magnitude to a low of about 9th, a brightness factor of 250 times. To help you make accurate estimates of the changing magnitude of Mira, Autumn Sky Window 5 shows the magnitudes of several nearby stars. These stars do not fluctuate in brightness, so they can be used as benchmarks for judging Mira. The magnitude of each comparison star is shown in parentheses, with each decimal point omitted to avoid confusing it with another star. Delta Ceti, for instance, is labeled "(41)," since it shines at magnitude 4.1, and so on.

**Through a Telescope.** Although Mira is bright enough to be seen with binoculars for most of its cycle, a telescope will be needed to spot it when near minimum. Mira's reddish color also becomes more pronounced when viewed through a telescope, especially if you defocus it slightly.

The name Mira means "wonderful," an apt name indeed for such a fascinating star. And what makes it so wonderful? Its fascinating behavior! With great regularity, Mira oscillates between 3rd and 9th magnitudes across 332 days, although it has brightened to 2nd magnitude and dimmed to 10th magnitude on rare occasions. In fact, once in 1779, it even brightened to 1st magnitude, outshining anything in the immediate area. But don't be deceived; Mira spends about 75 percent of the time below naked-eye visibility.

As mentioned earlier, an unrelated 9th-magnitude star lies just 2 arcminutes east of Mira and might actually outshine the variable when Mira is near minimum brightness. Both appear reddish through telescopes, which only adds to an observer's confusion. To confirm which is which, visit the Web site of the American Association of Variable Star Observers (www.aavso.org), where you will find detailed charts showing Mira and its immediate surroundings.

### Mira

Mira, first detected as a variable in 1596 by the German astronomer David Fabricius, is the archetypal long-period variable star. Long-period variables are symmetry in motion, as they slowly and predictably fluctuate between brightness maxima and minima. All are thought to be ancient red giants that actually pulsate over time due to internal changes. At maximum brightness, these stars expand to many times their minimum-brightness diameter. While astronomers believe these changes occur because the stars have spent the hydrogen fuel in their cores and are beginning to burn helium, the mechanism is not completely understood.

# 9

# Winter Sky Windows

Winter is a wonderful time for stargazing, with more bright stars adorning the sky (Figure 9.1) than any other season. Standing center stage is Orion the Hunter, the most brilliant constellation of all. Look for three equally bright stars in a row, forming his belt. To the northeast lies the bright reddish star Betelgeuse, the right shoulder of the Hunter. His left shoulder is represented by the star Bellatrix. A faint group of three stars depicting Orion's tiny head is centered above (north of) the shoulders. South of the belt are the stars Rigel, representing Orion's left knee, and Saiph, marking his right knee. He is usually depicted holding a shield of faint stars in his left hand, and a club raised high over his head in his right.

Orion is the best constellation for demonstrating that stars come in different colors, which indicate their temperatures. Most people are aware that flames of different temperatures glow in different colors. A yellow candle flame, for instance, is cooler than the blue flame of a gas stove. The same is true of stars, although I should quickly add that stars do not "burn," nor are they "on fire." The Sun is a yellow star and has a surface temperature of about 6,400 kelvin (11,000° Fahrenheit). Compare that to blue-white Rigel, one of the hottest stars known. Its surface temperature is believed to be 20,000 K (35,500° F). At the other end of the scale is red Betelgeuse, which has a surface temperature of only 3,000 K (5,000° F).

Though you can't tell by just looking at it with your eyes, Betelgeuse is one of the largest stars ever discovered. Classified as a red supergiant star, Betelgeuse measures about 640 million miles (1,030 million km) in diameter, more than 700 times larger than the Sun. Put another way, if Betelgeuse were at the center of our solar system, its outer edge would extend beyond the orbit of Mars. Earth would be inside the star!

# Winter sky

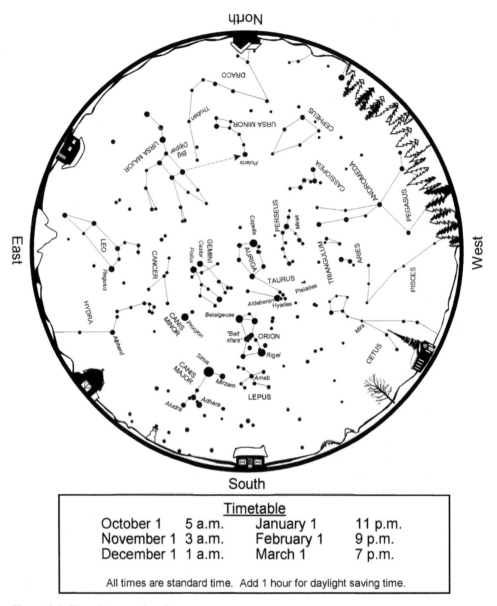

**Figure 9.1** *The winter evening sky.*

Hanging from the belt are the dim stars of Orion's sword. It turns out, however, that the middle star in the sword isn't a star at all but rather a cloud of glowing hydrogen gas called a nebula. Astronomers have cataloged this nebula as "M42," but most backyard stargazers know it better as the Great Orion Nebula. Read more about this nebula in Winter Sky Window 2.

According to legend, Orion was the mightiest hunter of all time. One day he boasted of being able to defeat any animal on Earth. His constant bragging was overheard by Mother Earth who, fearing that he would destroy every creature, sent a poisonous scorpion to sting Orion on the heel and kill him. Diana, goddess of the hunt, felt sorry for Orion. To honor him, she placed Orion among the stars. (The scorpion, Scorpius, was also placed in the sky. It may be found under heavy guard in the summer sky, directly opposite Orion in the sky so that it would never harm him again.)

It seems that there is no rest for the weary. Even in the sky, we find Orion doing battle again, this time with Taurus the Bull. To locate Taurus, draw a line through Orion's belt and extend it to the upper right (northwest). You will come to the bright orange star Aldebaran, marking the Bull's eye. The head of Taurus is formed by a V-shaped group of stars nicknamed the Hyades. Imaginary lines extend to two stars above its head to create Taurus's long horns.

Orion is depicted in the sky as trying to club the Bull over the head in order to save seven sisters who were kidnapped by Taurus. We can still see the sisters trapped in the sky, formed by the tiny cluster of stars known as the Pleiades. Most people can spot six or seven stars in this region, though on an extremely clear night, some can see up to fifteen. A telescope or binoculars will reveal that nearly a hundred stars form the Pleiades cluster, as is discussed below in Winter Sky Window 1.

Returning to the belt of Orion, continue the line in the opposite direction down toward the southern horizon to spot Sirius, brightest star of the nighttime sky. Sirius is also known as the Dog Star, as it belongs to the constellation Canis Major, the Large Dog. Can you see a dog among its stars? Take Sirius as a jewel on the dog's collar. The star Mirzam marks a front paw, Adhara a hind paw, and Aludra the tip of the hound's tail.

Canis Major is but one of Orion's two faithful companions. His small dog, Canis Minor, can be found to the north. Canis Minor is made up of only two easily visible stars, the brighter called Procyon. Most people can't actually see a dog here, except perhaps a hot dog!

Watching Orion and Taurus battle are the twin brothers Gemini. Their heads are marked by the bright stars Castor and Pollux, with their stick-figure-like bodies extending southwestward, standing on the hazy band of our Milky Way galaxy. Not as bright as in the summer, the winter Milky Way nonetheless contains many fascinating sights when searched with binoculars or a telescope. In ancient mythology, Castor was a famous horse trainer and soldier, while Pollux is remembered as Sparta's leading boxer.

Following the Milky Way northward, we find the constellation Auriga the Charioteer. The bright star Capella marks his left shoulder. Curiously, even though Auriga is referred to as a charioteer, he is frequently drawn as a shepherd holding three young goats, or kids. The major stars in Auriga form a celestial home plate or pentagon.

Some lesser known constellations also populate the winter sky. For instance, directly below Orion is the faint but easy-to-spot constellation Lepus the Hare. The body of the Hare is formed from a trapezoid of four stars, with the star Arneb marking its nose. Two stars form a northward-curving arc with

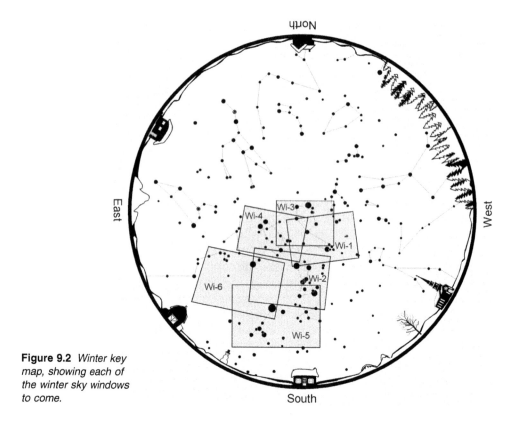

**Figure 9.2** *Winter key map, showing each of the winter sky windows to come.*

Arneb to create the animal's long ears, while two other stars to the west of the trapezoid could be seen as the ends of a carrot dangling in front of Lepus.

Figure 9.2 holds the key to the winter sky windows. The map, a duplicate of Figure 9.1, shows seven highlighted areas that are discussed in detail below. Each of those windows shows an enlarged area of the sky plotting fainter stars and selected deep-sky objects. Compare the two maps, decide which area you wish to explore tonight, and then head outside to find that region. Once the naked-eye stars and constellations have been located, switch to the close-up sky window found later in this chapter and, using your finderscope or binoculars, begin your journey into the depths of the universe.

Each sky window includes a miniature all-sky chart to show its position relative to the rest of the night sky. To help readers better appreciate the scale of things, a crosshair frame is also included in each window to show the typical field of a 6 × 30 finderscope (about 7°).

## Winter Sky Window 1

The first Winter Sky Window (Figure 9.3) includes three famous deep-sky objects: the Pleiades, the Hyades, and the Crab Nebula. All are found within the conspicuous constellation Taurus the Bull, highlighted by the orangish star Aldebaran.

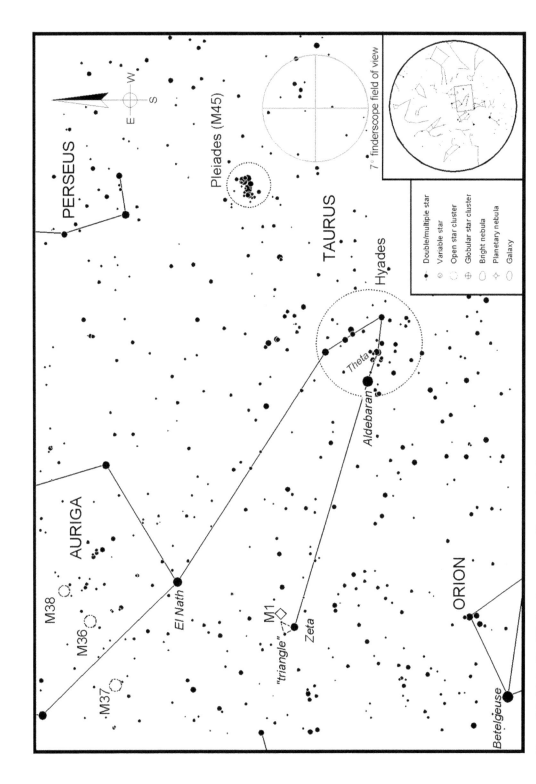

**Figure 9.3** *Winter Sky Window 1 opens onto Taurus the Bull.*

## Hyades: Open Cluster in Taurus

DISTANCE FROM EARTH: 151 light-years
FINDING FACTOR: *
"WOW!" FACTOR:    Binoculars: ****    Small telescopes (3″ to 5″): *
                 Medium telescopes (6″ to 8″): *

### Where to Look

The Hyades star cluster is very easy to find, as it marks the V-shaped head of Taurus the Bull. The bright star Aldebaran, although not a member of the Hyades cluster, can be seen even from heavily light-polluted areas.

### What You'll See

**Through Binoculars.** The Hyades is just wonderful through low-power binoculars! Most 7× glasses can squeeze it all into a single field of view, although higher powers probably cannot. Hundreds of stars are held within the group's 5.5° diameter, with more than 130 of them brighter than 9th magnitude and therefore visible in binoculars. For low-power, wide-field glasses, the Hyades ranks as one of the northern winter sky's best clusters.

Since Aldebaran is not a true member of the Hyades, the cluster's brightest star, at magnitude 3.4, is Theta-2 Tauri. Theta-2 teams with magnitude 3.8 Theta-1 Tauri to form a wide, naked-eye binary star. Both can be seen as part of the V pattern, just to the southwest of Aldebaran.

There is much more to the Hyades than just that V, however. Many of the stars scattered to its south belong to the cluster as well, including a prominent triangular pattern of three stars just to the south of Aldebaran. Another triangle of stars lies on the cluster's southwestern edge, while lanes of faint stars wind their way toward the northern edge.

**Through a Telescope.** As exciting as the Hyades are through binoculars and finderscopes, its impact is lost through the narrow fields of most telescopes.

### The Hyades

At 151 light-years away, the Hyades is one of the closest open star clusters to our solar system. Most of the stars in the Hyades are gathered into a central area 10 light-years across, although some stragglers are spread out over 80 light-years. Studies of the stars' spectra indicate the Hyades were formed some 660 million years ago.

The brilliant orange star Aldebaran is much closer to us than the Hyades, only about 60 light-years away.

## M45: Open Cluster in Taurus

NICKNAME: Pleiades, Seven Sisters, or Subaru
DISTANCE FROM EARTH: 380 light-years
FINDING FACTOR: *
"WOW!" FACTOR:    Binoculars: ****    Small telescopes (3″ to 5″): ***
                 Medium telescopes (6″ to 8″): ***

### Where to Look

The Pleiades is visible to the unaided eye as a tiny congregation of stars due west of Taurus's V-shaped head. Through light-polluted skies, you may just be able to make out their collective existence as a small fuzzy patch of light, while from darker skies, it becomes clear that the Pleiades is a collection of several faint stars. Most observers can count six set in a tiny dipper pattern, but given rural skies and good eyes, that number can grow to seven or even more.

### What You'll See

**Through Binoculars.** Is there any star cluster more spectacular through binoculars than the Pleiades? If so, I don't know of it. One look at them through 7× or 10× binoculars and you will feel like you've broken into the vault of that great jewelry store in the sky! There before our eyes, a dozen or more stellar sapphires, set ablaze against a velvety black background, shimmer with iridescence. Speckles of diamond dust, formed from fainter cluster stars, surround them.

Several striking binary and multiple stars highlight M45 (Figure 9.4). The stars Atlas, shining at magnitude 3.7, and Pleione, which varies in brightness from magnitude 4.8 to 5.7, form a wide pair that marks the eastern-pointing "handle" of the Pleiades' "bowl." Asterope is a wide pair of stars, while Alcyone, the brightest Pleiad, is a quadruple star system.

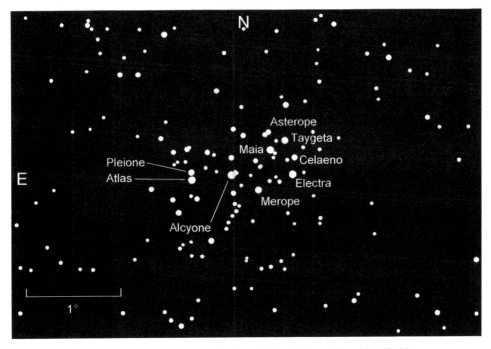

**Figure 9.4** *The Pleiades, long a binocular favorite, with its brightest stars identified by name.*

**Through a Telescope.** Due to its wide span covering four Full-Moon diameters, the Pleiades demand low power and a wide field to be seen at their best. Even a 2-inch (5-cm) telescope will cause a population explosion in the Seven Sisters' family by revealing dozens upon dozens of fainter stellar siblings. If the cluster proves too wide to squeeze into view, scan back and forth from one side to the other to drink in the entire sight.

### The Pleiades

Known since prehistoric times, the Pleiades have played different roles in different civilizations. While we call them the Pleiades or Seven Sisters, to the Japanese, they are Subaru. In fact, the Japanese auto manufacturer of the same name uses a seven-star pattern as their logo.

The Pleiades cluster has thrived in the cosmos for about 100 million years. Modern measurements show that it contains some five hundred stars spread out across 2°, with most too faint and scattered to be visible in amateur telescopes. Because of this weak density, astronomers expect the stars' mutual gravitational forces will hold the cluster together for another 250 million years or so, after which time the stars will disband.

Photographs, such as Figure 9.5, reveal an amazingly intricate network of bluish reflection nebulosity intertwined around the stars of M45. For years it was naturally assumed that these clouds were the leftover remains of the nebula that begat the cluster. Recent studies, however, prove that these clouds are an independent phenomenon that just happens to be passing through the same area of space at this time. Looks can certainly be deceiving!

**Figure 9.5** *The Pleiades, M45, and its associated reflection nebulosity are shown in this photograph taken by George Viscome through an 8-inch f/5.6 reflector. North is up.*

If you have an exceptionally clear evening, look closely and you just might glimpse soft gossamer wisps surrounding some of the brighter cluster stars. These gentle clouds belong to a large cloud of interstellar dust that happens to be passing through the Pleiades at this point in time. The brightest portion of the Pleiades nebulosity is identified as NGC 1435 and is found around Merope, the southeastern star in the Pleiades' bowl. With excellent sky conditions, some observers can make out its form in binoculars. Through 6- to 8-inch (15- to 20-cm) telescopes, it appears as a comet-shaped glow extending southward from the star, becoming more apparent as aperture increases. But don't be fooled by a haze that surrounds all of the Pleiades, an effect likely caused by dust or fog on your telescope optics. To find out if you are actually seeing the dim glow of the Pleiades nebula, take aim at the Hyades. If they are also engulfed in clouds, then your telescope optics are probably causing the effect, but if not, then you just might be seeing the real thing.

## M1: Diffuse Nebula in Taurus

NICKNAME: Crab Nebula
DISTANCE FROM EARTH: 6,300 light-years
FINDING FACTOR: **
"WOW!" FACTOR:  Binoculars: *       Small telescopes (3″ to 5″): **
                Medium telescopes (6″ to 8″): **

### Where to Look

From Aldebaran and the Hyades, trace the two horns of Taurus as they stretch northward toward the constellation Auriga. Center your telescope on the tip of the eastern horn, marked by the star Zeta Tauri, then look through the finderscope for two fainter stars to its north that join in to form a small triangle. Follow an imaginary line connecting these two suns about 0.5° to the west, where M1 awaits.

### What You'll See

**Through Binoculars.** I can still remember the cold January night back in high school when I first spotted the Crab through a pair of old 7 × 35 binoculars. I considered it to be a reasonably easy catch using my 8-inch (20-cm) reflector but impossible through binoculars, especially a pair that small. One night, when I was out in my backyard, lying in a snowbank and scanning back and forth through the winter stars, I decided to see how many Messier objects I could see. I thought it appropriate to start at the beginning, so I aimed toward M1. Sure enough, it was there! It was far from bright, but I could definitely make out a tiny smudge right where I knew the Crab lurked. You can do it too, but wait for an exceptionally clear night.

**Through a Telescope.** A 4-inch (10-cm) telescope shows the Crab Nebula (Figure 9.6) as a uniformly gray, oval glow. Many first-time visitors to M1 often comment on the nebula's unimpressiveness, so don't feel bad if it leaves you wanting more. In a way, that's good, because the next time you stop by M1,

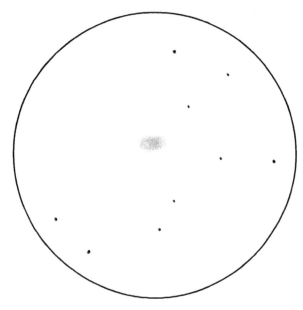

**Figure 9.6** *The Crab Nebula, M1, as drawn through the author's 8-inch reflector at 120×. South is up.*

## Crab Nebula

Nicknamed the Crab Nebula for its many spiny nebulous projections most easily visible in photographs, M1 is one of the most fascinating and mysterious objects found anywhere in the heavens. Since its discovery in 1731 by the London physician John Bevis, M1 has spurred on more interest and controversy than almost any other celestial object. When he first independently found it on August 28, 1758, Charles Messier mistook it for a comet. This singularly important observation ultimately led to the compilation of his now-famous list of deep-sky objects.

The Crab's impact on our view of the universe began long before that, however. On July 4, 1054, Chinese astronomers studying the early predawn sky suddenly noticed a brilliant star where no star previously existed. It outshone the planet Venus, and even though the Sun was nearby at the time, it was visible in broad daylight for more than three weeks. Halfway around the globe, drawings attrib-

uted to Anasazi and Mimbre Native American artists also captured its appearance. Strangely, there are no records that astronomers in Europe, just emerging from the Dark Ages, ever noticed the event.

Today, we know that they all witnessed the final hurrah of an extremely massive star. As mentioned in chapter 1, when the universe's most massive stars reach the end of their lives, they become very unstable. As enormous temperatures and pressures develop, the star's iron-rich core collapses rapidly, triggering an all-consuming explosion called a supernova. All that is left is an expanding cloud of gaseous debris and, buried deep within, the rapidly beating heart of that ancient star: the Crab Nebula pulsar, the crushed remains of the star's core. This incredibly dense stellar oddity rotates thirty times each second, sweeping a beam of energy across Earth with every pass. The Crab pulsar remains one of the fastest of its kind to this day.

you may notice one or two subtle details that you missed before. Don't forget to use averted vision.

The intricate irregularities that are so prominent in photographs don't begin to appear until viewed through 10-inch (25-cm) and larger instruments. Large apertures increase the mottled look of M1, although the famous crab-like appearance is difficult to detect visually.

## Winter Sky Window 2

No other constellation in the sky draws people's attention like Orion the Hunter, framed in Winter Sky Window 2 (Figure 9.7). Standing boldly above the southern horizon during the winter months, the Hunter is made prominent by the three equally spaced stars we call his belt and four brilliant suns framing his body. All are visible from just about anywhere on Earth. As impressive as Orion is, his telescopic treasures, and those of the surrounding region, are even more exciting.

### M42 and M43: Bright Nebulae in Orion

NICKNAME: Orion Nebula
DISTANCE FROM EARTH: 1,600 light-years
FINDING FACTOR: *
"WOW!" FACTOR:    Binoculars: ****    Small telescopes (3″ to 5″): ****
Medium telescopes (6″ to 8″): ****

#### Where to Look

Begin by locating Orion's Belt. Below those three stars, look for a small group of two or three very faint stars, called Orion's Sword. Aim your telescope at these stars. Through your finderscope, you may see that the middle star in the Sword is surrounded by a fuzzy glow. That glow is our target, M42, the Orion Nebula.

#### What You'll See

**Through Binoculars.** M42 puts on a magnificent show through all telescopes and binoculars. Through 6× and 7× binoculars, as through a finderscope, the nebula appears as a faint, cloudy glow surrounding two stars, known to astronomers as Theta-1 and Theta-2 Orionis. If the nebula is tough to pick out at first, especially from light-polluted skies, try using averted vision. From darker skies, the cloud takes on an irregular, curved shape surrounding those stars.

**Through a Telescope.** Begin your visit using your lowest magnification, between 40× and 60×. The nebula will look like a soft glow engulfing the stars Theta-1 and Theta-2 Orionis (Figure 9.8). A 3-inch (7.5-cm) telescope will begin to show some of the intertwined irregularities that characterize the true, complex structure of the Orion Nebula. Theta-1 Orionis sits near the tip of what looks like the silhouette of a finger sticking into M42. Called the Fish's Mouth,

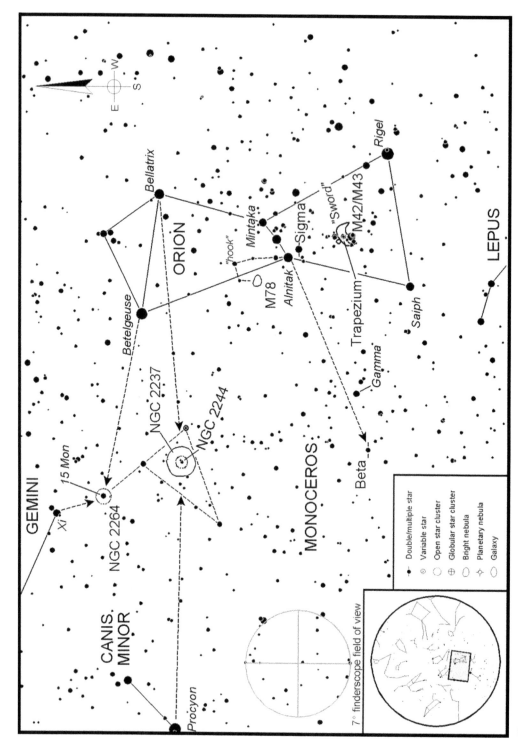

**Figure 9.7** *Winter Sky Window 2 holds a whole host of spectacular deep-sky objects, including a perennial favorite, M42.*

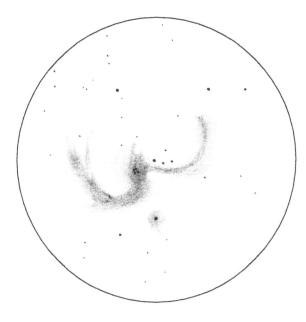

**Figure 9.8** *M42 and M43, as drawn through the author's 8-inch reflector at 55×.*

## Orion Nebula

The Orion Nebula, a huge cloud of hydrogen gas and dust, is the sky's best-known example of an emission nebula or Hydrogen-II region (Figure 9.9). Think of it as a stellar maternity ward. Hidden among the tufts and eddies of nebulosity, protostars are rotating and compressing as they continue on the road to stardom. The stars of the Trapezium are among the youngest, hottest suns visible through backyard telescopes, but there are more stars within the Orion Nebula than meet the eye. Buried within, a cluster of a thousand stars remains obscured behind clouds of opaque dust. They are only visible in infrared images, which can penetrate the dust.

Ultraviolet radiation from these hot, newly born stars in turn *ionizes* the surrounding atoms of hydrogen gas by breaking the bonds between the atoms' nuclei and the circling electrons. When hydrogen is ionized, the neutral atom is split into a positive hydrogen ion (a proton) and a free electron. That free electron won't stay free for long, however. Instead,

it will be captured by another hydrogen ion to form neutral hydrogen once again. In the process, the hydrogen gas emits red (called hydrogen-alpha) and green (hydrogen-beta) light. In a sense, M42 is a cosmic neon sign, since the process that causes it to glow is very similar to what happens as electrical current passes through a neon sign here on Earth.

Color photographs show the Orion Nebula glowing the brilliant red of ionized hydrogen. Unfortunately, there are many telescope companies out there that, through misleading advertising, try to fool consumers into thinking this is also the view they will see through their telescopes. When we look toward M42 with our telescopes, that strong coloring is nowhere to be seen. It's a sad but true fact that the human eye is nearly color blind under dim light, especially to the color red (that's why we use red flashlights!). Given crystalline skies, the best that observers with 6-inch (15-cm) and larger telescopes will be able to see is a greenish tint.

**Figure 9.9** *South is toward the top of this magnificent photograph of the Great Nebula in Orion, M42, taken by George Viscome through a 14.5-inch f/6 reflector. M43 appears as a hook-shaped appendage to the north of (below) the main nebula.*

this formation is caused by a dark cloud set in front of the brighter area that lies beyond.

Now switch to a 100× eyepiece and concentrate your view on the region of the Fish's Mouth and Theta-1 Orionis. Theta-1 is not a single star, as it may have appeared at low power but instead is actually a family of four suns neatly gathered in a trapezoid, called the Trapezium. Spotting all four stars through a 3-inch (7.5-cm) is a fun way to test the telescope's optical quality. Four- and 6-inch (10- and 15-cm) telescopes should show all four stars readily. The Trapezium's stars are designated with letters (A, B, C, and D), ordered according to their location. The system's primary star, the brightest of the bunch at magnitude 5.1, is known as Theta-1C and marks the trapezoid's southern corner. The western star (Theta-1A) and the northern star (Theta-1B) are both known to be eclipsing binaries, with a smaller companion star alternately passing in front of and behind the larger primary star. These eclipses cause both stars to vary slightly in brightness, though they usually shine at magnitude 6.7 and 7.9, respectively. Theta-1D also shines at magnitude 6.7.

If the seeing is steady, take a look at the Trapezium with 150× to 200×. You just might be able to spot two additional, fainter stars. Theta-1E glows dimly at 11th magnitude and is only 4 arc-seconds northwest from the A star. Some people may see it as slightly orange. Theta-1F is about half a magnitude fainter and hides only 4 arc-seconds from the C star. Spotting these two hidden treasures is a real test of your visual acuity, but don't be surprised or disappointed if you don't see them at first. It took me *years* to pick them out with my 8-inch (20-cm) reflector, though I have subsequently seen them in a 4-inch (10-cm) refractor.

Six-inch (15-cm) and larger telescopes add many faint tendrils of nebulosity that hook away from the Fish's Mouth and curve toward the southeast and northwest. The overall appearance always reminds me of a cupped hand, with the "thumb" extending toward the west, the fingers toward the east, and the Trapezium held in the palm. A narrowband nebula filter or an Oxygen-III filter can greatly enhance these dim extensions, especially from urban and suburban observing sites.

Just north and slightly east of M42 is a second, much smaller tuft of nebulosity cataloged separately as M43. Two- and 3-inch (5- to 7.5-cm) telescopes easily show its glow, while 6- and 8-inch (15- and 20-cm) telescopes reveal that M43 is not perfectly symmetrical but shaped a little like a comma hooking toward the north. In reality, M43 is an extension of the Orion Nebula and only looks "separated" because of a tuft of dark nebulosity that slices between the two.

## M78: Bright Nebula in Orion

DISTANCE FROM EARTH: 1,600 light-years
FINDING FACTOR: **
"WOW!" FACTOR:    Binoculars: *     Small telescopes (3″ to 5″): **
                 Medium telescopes (6″ to 8″): ***

### Where to Look

M78 lies about half a finder field east of the star Mintaka, the northwestern star in Orion's Belt. Thanks to that coincidence, one way to find M78 is to have Earth do it for you. By aiming at Mintaka with your lowest-power eyepiece and letting the stars drift through the field for fourteen minutes, Earth's rotation will bring M78 into view. Another way to find M78 is to aim at Alnitak, the southeastern star in the Belt, and trace out the hook pattern of four stars shown in Winter Sky Window 2. M78 might just be bright enough to see through your finderscope as a weak glow.

### What You'll See

**Through Binoculars.** Binoculars, like finderscopes, are just capable of showing M78, given halfway decent skies. Look for a small, soft tuft of cloudiness. Two dim stars embedded within the cloud might also be spotted through 70-mm or larger binoculars.

**Through a Telescope.** M78 stands out well in small telescopes, with most showing what looks like a comet (Figure 9.10). Two faint "nuclei" can be imagined within the "comet's coma," while a broad, but short "tail" extends to their south. It turns out that those "nuclei" are actually a pair of nearly identical 10th-magnitude stars that are responsible for lighting the nebula.

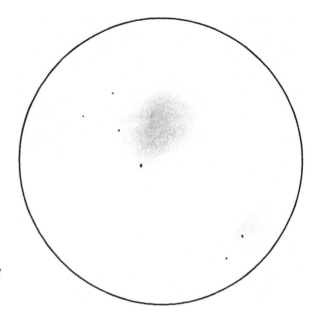

**Figure 9.10** *Reflection nebula M78 in Orion, drawn through the author's 8-inch reflector at 120×. NGC 2071 appears as a fainter glow to the northeast (lower right in the drawing, which is oriented with south toward the top).*

### Reflection Nebula in Orion

Discovered by Pierre Mechain in 1780 and added to Messier's catalog later that same year, M78 is part of the same huge molecular cloud as M42, and so is about the same distance away. But while M42 actually glows under its own power, M78 only reflects the light from nearby stars. Although there is hydrogen gas in M78 as well, the energy imparted by those two embedded stars is not enough to cause it to glow. Instead, starlight reflects off of grains of interstellar dust that are in the cloud, creating what astronomers call a *reflection nebula*. Reflection nebulae are always easy to identify in photographs by their deep-blue color, caused by the dust grains scattering blue light. This is similar to the scattering process that makes our daytime sky blue.

## Sigma Orionis: Multiple Star in Orion

DISTANCE FROM EARTH: About 820 light-years
FINDING FACTOR:  *
"WOW!" FACTOR:    Binoculars: not resolvable    Small telescopes (3″ to 5″):  ***
                 Medium telescopes (6″ to 8″):  ***

### Where to Look

Sigma Orionis is easy to find just 1° southwest of Alnitak, the southeastern star in Orion's Belt. Put Alnitak in the center of your finderscope, then look for a slightly dimmer star to the southwest. That's Sigma.

### What You'll See

**Through Binoculars.** I am sorry to say that while Sigma is bright enough to be seen with the naked eye, its trio of stars are too close to be separable through binoculars.

**Through a Telescope.** Even with only a modest, 2-inch (5-cm) telescope, Sigma Orionis will show itself to be more than a single point. The brightest star in the system is labeled Sigma A, and although it is a close double star, its nearby companion, Sigma B, is too close to be easily separable through most backyard telescopes. Sigma C, a 10th-magnitude sun, is also too close to Sigma-A to be easily seen, but two other members of the system are readily apparent even with low magnification. Seventh-magnitude Sigma D can be found just 14 arc-seconds to the east, while 6.5-magnitude Sigma E lies 41 arc-seconds northeast. All three create a slender triangular pattern, glistening white or perhaps slightly bluish through backyard telescopes.

#### Sigma Orionis

Sigma Orionis is actually a five-star (quintuple) system. Sigma A and B are about 100 astronomical units (9.3 billion miles or nearly 15 billion km) apart! Even though that's more than twice as far as Pluto is from our Sun, it is still too close for the two stars to be separable through backyard telescopes. Both are extremely hot, bright blue giant stars, emitting many times the energy of our Sun. Sigma D is the faintest of the system's three visible stars. Some observers see it as reddish, although that faint tint misses me. It lies some 4,500 astronomical units (418.5 billion miles or 674 billion km) from A-B. Finally, Sigma E, another blue giant, is nearly 2 trillion miles (3.2 trillion km) from Sigma A-B.

## Beta Monocerotis: Multiple Star in Monoceros

DISTANCE FROM EARTH: 690 light-years
FINDING FACTOR: **
"WOW!" FACTOR:    Binoculars: not resolvable     Small telescopes (3″ to 5″): ***
                 Medium telescopes (6″ to 8″): ***

### Where to Look

Beta Monocerotis is isolated in the dim constellation Monoceros the Unicorn, far from the bright stars of the winter. The easiest way to find it is to use the stars of Orion's Belt. Through your finderscope, trace a line as you would from the Belt stars toward Sirius in Canis Major, but hook slightly to the east instead of going straight to that brilliant star. That should bring you to Beta, as shown on Winter Sky Window 2. Take care not to confuse Beta with Gamma Monocerotis to its west, which is actually slightly brighter. Gamma is surrounded by several stars that should be visible in your finder, while Beta appears more isolated.

## What You'll See

**Through Binoculars.** Binoculars can be very helpful for locating Beta in the sky, since it might be too faint to be seen readily with the eye alone through light-polluted skies. However, the star's components are too close to one another to be resolved with ordinary binoculars.

**Through a Telescope.** Beta Monocerotis is one of the sky's premier triple stars. What looks like a single point visible through finderscopes and binoculars is actually a brilliant stellar triple play. The system's "A" sun is separated from the slightly fainter "B" star by about 7 arc-seconds, while the dimmer "C" companion lies slightly farther away still. All three appear dazzlingly white against a star-filled vault.

### Beta Monocerotis

The Beta Monocerotis system is formed from three nearly identical type B suns in orbit about one another. The brightest of the three, christened "Beta Mon A," shines at magnitude 4.7. The nearby "B" star shines at magnitude 5.2, while the more distant "C" star is magnitude 6.2. The distance between the two brightest components is believed to be about 1,600 astronomical units.

Six-inch (15-cm) and larger telescopes may also show a dim fourth star in the Beta family. The feeble light of this 12th-magnitude sun raises the system's population to four. Look for this challenging star about 26 arc-seconds to Beta-A's northeast.

## NGC 2244: Open Cluster in Monoceros

DISTANCE FROM EARTH: 5,500 light-years
FINDING FACTOR: **
"WOW!" FACTOR:    Binoculars: **        Small telescopes (3″ to 5″): ***
                 Medium telescopes (6″ to 8″): **

### Where to Look

Although open cluster NGC 2244 is a bright collection of stars, locating it inside the obscure constellation Monoceros often frustrates observers. Monoceros is the "hole" in the winter sky framed by Betelgeuse in Orion, Procyon in Canis Minor, and Sirius in Canis Major. To find NGC 2244, draw a line between the stars Procyon and Bellatrix, in Orion's left (western) shoulder. Aim your finderscope halfway along, seemingly at nothing. A look through the finder, however, should show NGC 2244 as a fuzzy blotch inside a triangle of 4th-magnitude stars that just squeezes into your finder field of view.

### What You'll See

**Through Binoculars.** The triangle mentioned above is also easily viewable through binoculars and makes a great landmark for finding NGC 2244. Binoculars reveal half a dozen 6th- and 7th-magnitude stars in NGC 2244, set in a distinctly rectangular pattern and highlighted by the yellowish glint of the 6th-magnitude star 12 Monocerotis.

**Through a Telescope.** With the extra magnification and light-gathering afforded by a telescope, NGC 2244 expands into a grouping of six bright stars set in a crooked rectangle and peppered by another twenty-five or so fainter suns. The brightest six stars really sparkle against the starry backdrop.

Try viewing NGC 2244 with your lowest-power eyepiece and see if you can detect a faint glow encircling the cluster. Nothing? If you have a light-pollution filter, screw it into place and give it another go. That glow is the gentle glimmer of NGC 2237, the Rosette Nebula, a dim wreath of nebulosity surrounding the open cluster. The Rosette measures more than a degree in diameter and is very tough to spot. I have seen it through a 4-inch (10-cm) telescope, but only when there was no appreciable light pollution.

### NGC 2244

Discovered in 1690 by John Flamsteed, NGC 2244 is one of the most attractive open clusters in the winter sky. Forty suns make up the group, which spans approximately 50 light-years.

NGC 2244 is a young open cluster, with stars probably no more than 4 million years old, all having formed from the surrounding Rosette Nebula. The energy from the cluster's young stars is ionizing the surrounding hydrogen gas clouds, causing them to glow a vivid red, while streams of hot particles have created a hole in the nebula's center to give it the wreathlike appearance seen in photographs.

## NGC 2264: Open Cluster in Monoceros

NICKNAME: Christmas Tree Cluster
DISTANCE FROM EARTH: 2,400 light-years
FINDING FACTOR: **
"WOW!" FACTOR:    Binoculars: **        Small telescopes (3″ to 5″): ***
                 Medium telescopes (6″ to 8″): ***

### Where to Look

There are a couple of ways to find NGC 2264. To me, the easiest is to draw a line from Bellatrix to Betelgeuse in Orion, then extend it about 1.5 times to the east. That will place you almost directly below the star Xi Geminorum, the foot of the twin brother Castor, in Gemini, where you should find a lone 5th-magnitude star, labeled 15 Monocerotis on Winter Sky Window 2. Aim there and have a look through your telescope, since 15 Mon is actually the brightest member of NGC 2264.

### What You'll See

**Through Binoculars.** Nicknamed the Christmas Tree Cluster for its distinctive stellar pattern, this large, bright open cluster is easily found through all binoculars. Its brighter stars fall into a conelike shape reminiscent of a seasonal evergreen.

**Through a Telescope.** NGC 2264 is a fun cluster to show friends and family. Center it in the field of your lowest-power eyepiece, then step back and let them have a look. Ask them if the stars in the cluster seem to draw a distinctive shape or pattern. At least one person will probably say a Christmas tree. The cluster's brightest star, 15 Monocerotis, marks the trunk, while the tree extends toward the south. Ten stars or so serve as lights at the ends of invisible branches. The Christmas tree shape is easy to pick out in small telescopes, but instruments larger than 8 inches (20 cm) in aperture may have a tough time seeing the shape, since their fields of view may be too narrow to fit it all in.

### Christmas Tree Cluster

In all, forty stars belong to NGC 2264. The cluster's brightest star, 15 Monocerotis, is also known as S Monocerotis, an irregular variable star that fluctuates ever-so-slightly between magnitudes 4.6 and 4.7.

Although it is nearly impossible to see visually except through a large telescope, a triangular wedge of dark nebulosity extends from the star at the top of the Christmas Tree. Known as the Cone Nebula for its appearance on long-exposure photographs, this interstellar cloud is only a small portion of a large complex of nebulosity where new stars are being formed.

## Winter Sky Window 3

Passing near the zenith on cold winter nights, the third winter sky window (Figure 9.11) opens onto the constellation Auriga the Charioteer. The bright star Capella lights the way, aiding in finding Auriga's pentagonal body.

## M36: Open Cluster in Auriga

DISTANCE FROM EARTH: 4,100 light-years
FINDING FACTOR: **
"WOW!" FACTOR:    Binoculars: **        Small telescopes (3″ to 5″): ***
                 Medium telescopes (6″ to 8″): **

### Where to Look

To find M36, center your finder about halfway between El Nath, the star that also marks the end of one of Taurus's horns, and Theta Aurigae, marking the Auriga pentagon's northeastern corner. Slide about a third of a finder field to the west-northwest, where you should see M36 as a dim patch of fuzzy light nestled in between two faint stars.

### What You'll See

**Through Binoculars.** The stars of M36 blend into a gentle stellar fog when viewed through most binoculars, although a few weak points of light might break through the mist.

**Through a Telescope.** M36 is a wonderful sight through all telescopes, especially when viewed at lower magnifications (Figure 1.2). The smallest backyard instruments should show about ten stars against a hazy backdrop, while

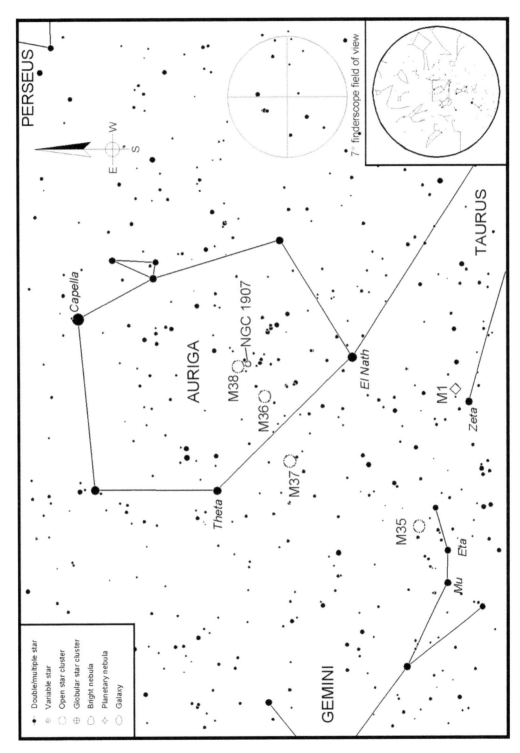

**Figure 9.11** *Winter Sky Window 3 centers on Auriga the Charioteer.*

an 8-inch (20-cm) will up the star count to more than three dozen points that seem to form a crooked Y pattern.

Just south of the cluster's center lies the double star Struve 737 (usually abbreviated Σ737 in references). Switch to a medium magnification to make out its two 9th-magnitude suns separated by about 10 arc-seconds.

---

### M36

M36, discovered by Giovanni Hodierna sometime before 1654, holds sixty stars, mostly hot, spectral-type B suns, within its 14-light-year gravitational grip. Studies of the light from these stars show that many are rotating rapidly, a trait also seen in other star clusters, such as the Pleiades. Judging by the lack of any red giant stars, M36 is thought to be fairly young, probably about 25 million years old. This makes it younger than nearby M37 and M38, which lie roughly the same distance away.

---

## M37: Open Cluster in Auriga

DISTANCE FROM EARTH: 4,400 light-years

FINDING FACTOR: **

"WOW!" FACTOR:     Binoculars: **          Small telescopes (3″ to 5″): **

                  Medium telescopes (6″ to 8″): ***

### Where to Look

Just as M36 was west-northwest of the halfway point between the stars El Nath and Theta Aurigae, M37 lies an equal distance to the east-southeast. It will appear dimmer through your finderscope than its neighbor to the west, but it is still bright enough to be visible.

### M37

Also discovered by Giovanni Hodierna sometime before 1654, M37 contains about 150 stars brighter than 13th magnitude and perhaps as many as 500 stars altogether. They are all crammed into an area that only measures about 25 light-years across, which makes M37 one of the denser open clusters in the winter sky.

### What You'll See

**Through Binoculars.** Since the brightest stars in M37 are about 9th magnitude, below the limit for most binoculars, M37 is usually seen as a starless patch of misty light. If skies are very dark, you might catch a glimpse of a couple of dim specks peering out through the fog.

**Through a Telescope.** Although its stars are dimmer than those in M36, M37 is the more striking open cluster in my opinion. Smaller scopes, like binoculars, will display a grainy, oval patch of starlight with perhaps four or five weak specks of light buried within. The brightest star staring back at you, a 9th-magnitude orb, might show an orangish cast if you defocus its image slightly. The misty appearance begins to dissolve into separate stars with a 6-inch (15-cm) telescope, with most shining between 11th and 13th magnitude (Figure 9.12).

**Figure 9.12** *M37, one of three Messier open star clusters in Auriga. South is up in this photograph by George Viscome taken through a 14.5-inch f/6 reflector.*

## M38: Open Cluster in Auriga

DISTANCE FROM EARTH: 4,200 light-years
FINDING FACTOR: **
"WOW!" FACTOR:    Binoculars: **        Small telescopes (3" to 5"): ***
Medium telescopes (6" to 8"): ***

### Where to Look

M38 is located almost centrally inside the pentagonal body of Auriga, the Charioteer. From M36, move half a finder field to the northwest, where you should see M38's hazy presence near a small triangle of field stars.

### What You'll See

**Through Binoculars.** M38 is a very pleasant sight in binoculars. Most observers will see it as a circular glow with a few scattered faint stars. Take a careful look and you might see that the stars seem to be arranged in one, two, or three crisscrossing rows or columns.

**Through a Telescope.** With more cluster stars added to flesh out the pattern, M38's "linear" appearance is much more striking through telescopes than through binoculars. While the stars of M38 can be seen to form any of a number of different patterns, the most obvious to my eye is one that resembles the Greek letter $\pi$ (pi). Two parallel rows extend roughly north-south, while a top "bar" of stars bridges the gap between them.

Here's a bonus object! Take a look just half a degree to the south-southwest of M38 for a second, smaller patch of misty starlight. That's NGC 1907, a separate star cluster that was missed by Messier. A 4-inch (10-cm) telescope should be enough to show a few very faint individual stars within a round glow. Of the thirty stars that call NGC 1907 home, the brightest shines at 11th magnitude.

# Winter Sky Window 4

Marked by the bright stars Castor and Pollux, the constellation Gemini stands tall in Winter Sky Window 4 (Figure 9.13). Here we find an interesting multiple star, several star clusters, and even a planetary nebula.

## Castor: Multiple Star in Gemini

DISTANCE FROM EARTH: 52 light-years

FINDING FACTOR:  *

"WOW!" FACTOR:    Binoculars: not resolvable      Small telescopes (3″ to 5″):  **
                  Medium telescopes (6″ to 8″):  ***

### Where to Look

Castor is easy to find, even from a light-polluted city. Look for it and its "twin" Pollux to the north of Orion. As Earth slowly turns the stars at night, Castor, the more westerly of the two, leads Pollux across the sky.

### What You'll See

**Through Binoculars.** Although Castor appears as a brilliant whitish sun through binoculars, more magnification is required to see it as anything more than a single point of light.

**Through a Telescope.** Viewed through a 2.4- or 3-inch (6- to 7.5-cm) telescope, Castor is revealed to be a double star comprised of two brilliant 2nd- and 3rd-magnitude suns separated by 4 arc-seconds. A high-quality 4-inch (10-cm) telescope should be able to add a third, much fainter star to the mix. Castor C, as it is called, shines only at 10th magnitude and is separated from the others by about 73 arc-seconds.

> ### Castor
>
> Stars A and B are separated from each other by about 100 astronomical units (ten times the distance from the Sun to Pluto) and revolve about each other in approximately 470 years. Star C, lying some 1,200 astronomical units from the others, probably takes more than 10,000 years to complete an orbit.
>
> Although Castor appears to be a triple star visually, it turns out that there is more here than meets the eye. Both Castor A and Castor B have close-by companion stars, neither of which are visible directly. Astronomers have detected their presence, however, by the gravitational "wobble" they create in the lines that form the stars' spectra. Not to be left out, Castor C is an eclipsing binary, bringing the total number of stars in the system to six.

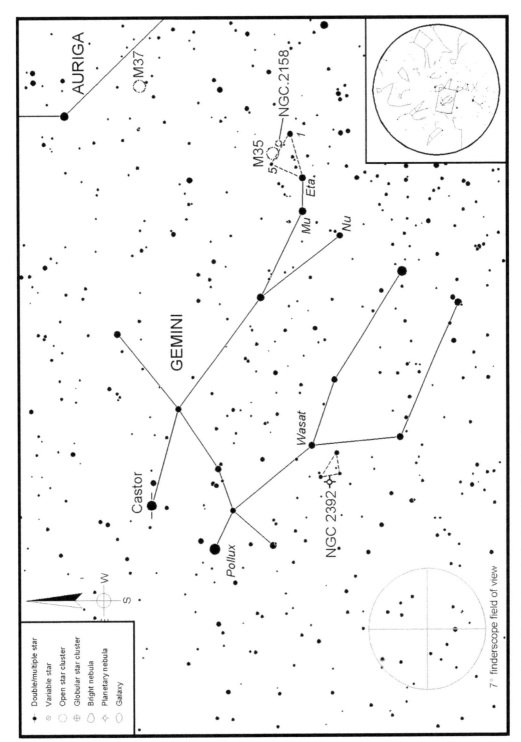

**Figure 9.13** *Winter Sky Window 4 is dominated by Gemini the Twins.*

## M35: Open Cluster in Gemini

DISTANCE FROM EARTH: 2,800 light-years

FINDING FACTOR: **

"WOW!" FACTOR:    Binoculars: ***    Small telescopes (3″ to 5″): ***
Medium telescopes (6″ to 8″): ***

### Where to Look

Begin by picking out the two stick-figure brothers, marked by the bright stars Castor and Pollux. Trace Castor's body toward the southwest, ending at the stars Mu and Nu Geminorum, which mark his feet. Aim your telescope or binoculars toward Mu. Through the finder, it appears that Castor's foot must be double-jointed, as Mu combines with the stars Eta and 1 Geminorum to form a three-starred arc. A fourth, fainter star, 5 Geminorum, creates a right triangle with 1 and Eta. M35 lies near 5 Gem, at the triangle's right angle.

### What You'll See

**Through Binoculars.** Under dark skies, M35 is just bright enough to be seen with the eye alone, and so should be visible through binoculars fairly readily under all but the most oppressive observing conditions. Most binoculars will reveal the brightest half dozen or so cluster members against the strong glow from other stars that are too faint to be resolved individually.

**Through a Telescope.** M35 (Figure 9.14) is dazzling even through small back-yard telescopes. Three- to 4-inch (7.5- to 10-cm) instruments show dozens of points of light within M35, accentuated by those half a dozen or so brighter suns. Most of the cluster stars are blue-white giants, although a few show tints of yellow and orange. Many form graceful arcs and curves that thread throughout the cluster, although there is a curious absence of stars near the center of the group. Be sure to use a low-power eyepiece for the best view.

Here's another bonus object. Just southwest of M35 is the very rich and very distant open cluster NGC 2158. While NGC 2158 as a whole is just bright enough to be visible through a 4-inch (10-cm) telescope, a 6- or 8-inch (15- or 20-cm) telescope must be used to see it as anything more than a vague glow, since its brightest stars are only 12th magnitude.

### M35

The discovery of M35 is credited to the Swiss astronomer Philippe Loys de Cheseaux, who spotted it in either 1745 or 1746. More than two hundred stars scattered across about 1° of sky populate M35, which extends about 24 light-years across. Several orange and yellow giant stars indicate that M35 is probably in excess of 100 million years old, a moderate age as open clusters go.

M35's neighbor, NGC 2158, is hardly neighborly in reality. In fact it lies nearly 16,000 light-years from us, close to the outer fringes of our galaxy.

**Figure 9.14** *M35 in Gemini, one of winter's finest open star clusters. The distant open star cluster NGC 2158 is seen in the upper left (southwest) corner of this photograph taken by George Viscome through a 14.5-inch f/6 reflector. South is up.*

## NGC 2392: Planetary Nebula in Gemini

NICKNAMES: Eskimo Nebula, Clown-Face Nebula
DISTANCE FROM EARTH: 5,000 light-years
FINDING FACTOR: **
"WOW!" FACTOR:    Binoculars: *        Small telescopes (3″ to 5″): **
Medium telescopes (6″ to 8″): ***

### Where to Look

Once you spot the constellation Gemini, zero in on the star Wasat, which marks Pollux's hip bone. Look through your finderscope about half a field to the southeast of Wasat for three 5th- and 6th-magnitude stars that form a right triangle. Set your aim on the star marking the triangle's northern corner. With a low-power eyepiece in place, shift about half a degree farther southeast, to find NGC 2392.

### What You'll See

**Through Binoculars.** NGC 2392 shines at 8th magnitude and spans nearly 13 arc-seconds, which is a winning combination for binoculars. Actually, through low-power glasses, NGC 2392 can be mistaken for the southern half of a double star, as it is found just south of an unrelated 8th-magnitude field star. What gives the planetary away is its tiny, vibrant blue disk, which is far more color-ful that any of the surrounding stars.

**Through a Telescope.** NGC 2392 also looks like a tiny blue disk in 2- and 3-inch (5- and 7.5-cm) telescopes, while 6-inch (15-cm) and larger instruments add the twinkle of its 11th-magnitude central star. When viewed at high magnification, NGC 2392 reveals a fuzzy outer region surrounding a brighter triangular center, as shown in Figure 9.15. William Herschel was the first to imagine the inner region as a human face, with the central star marking the nose, and the outer surrounding ring as a fur-lined hood or perhaps a clownlike ruffled collar. This fanciful view led NGC 2392 to be nicknamed the Eskimo or Clown-Face Nebula.

### Eskimo or Clown-Face Nebula

Discovered by William Herschel in 1787, NGC 2392 is one of the winter sky's brightest planetary nebulae, the expanding outer shell having been discharged from an aging star as the star underwent tremendous internal turmoil. We see the result of that stress today in the form of an expanding cloud surrounding the remains of the star, now a white dwarf. Astronomers place the age of the nebula at 10,000 years. Photos taken with the Hubble Space Telescope show spectacular streamers hurtling away from the cloud's brighter central portion. Portions of the nebula are believed to be expanding at an incredible 900,000 miles per hour (1.5 million km per hour)! NGC 2392 is estimated to measure about 1 light-year across.

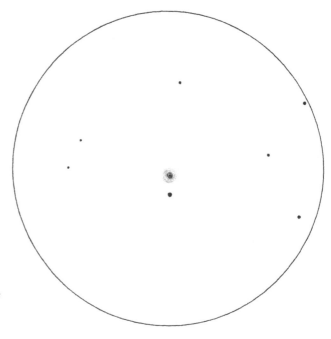

**Figure 9.15** *NGC 2392, the Eskimo or Clown-Face Nebula, in Gemini, as drawn through the author's 8-inch reflector at 203×. South is up.*

# Winter Sky Window 5

We leave the northern portions of the winter sky behind and move toward the south, where the constellation Canis Major is found running along the horizon, faithfully following its master, Orion. The brilliant beacon Sirius draws immediate attention, while the stars that surround it are considerably fainter and less obvious. Other constellations found within Winter Sky Window 5 (Figure 9.16) include Lepus the Hare, directly south of Orion, and portions of Puppis, the Stern, to the east of Sirius and Canis Major. Several spectacular Messier objects lie in wait in this region, but their low elevations can make spotting them difficult if trees or other earthly obstacles lie in this direction.

## M41: Open Cluster in Canis Major

DISTANCE FROM EARTH:  2,300 light-years

FINDING FACTOR:  *

"WOW!" FACTOR:    Binoculars: **       Small telescopes (3″ to 5″): ***
                 Medium telescopes (6″ to 8″): ***

### Where to Look

Open cluster M41 is located about half a finderscope field of view directly south of Sirius. In your mind, draw a right triangle from Sirius to the star Mirzam, often shown as depicting the Dog's front paw, and an invisible point to the south of Sirius. Aim your telescope at that spot and take a look for a hazy smudge of grayish light. That will be M41. In fact, from a dark-sky site, M41 might even be visible without any telescopic aid at all.

### What You'll See

**Through Binoculars.** This is a seasonal favorite open cluster for binoculars. Even 7× glasses are able to pick out around twenty stars within the fuzzy boundaries of M41, all sparkling like sapphires against black velvet.

**Through a Telescope.** M41 is a striking open cluster through telescopes. Even a 3-inch (7.5-cm) telescope will show some thirty to thirty-five individual stars set against the dim glow of still fainter suns. The brighter stars at the heart of M41 appear to form a trapezoidal "keystone" figure that is strongly reminiscent of the constellation Hercules. Also take a look at how many of the fainter cluster stars are set in closely tied pairs and triples.

If you have color-sensitive eyes, M41 will appear more colorful than most open clusters. One of the brighter stars in the center of M41 glows with a reddish tint, while some of the fainter members shine with hints of yellow and orange. Try slightly defocusing the image to make their colors a little more distinct.

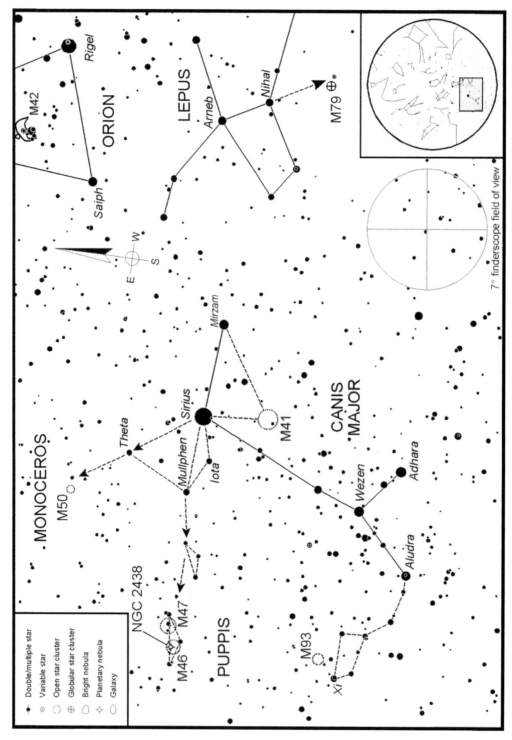

**Figure 9.16** *Winter Sky Window 5 includes the bright star Sirius, its home constellation Canis Major, and Lepus, all to the south of Orion.*

*M41*

Records show that M41 was discovered long before the invention of the telescope. In fact, the Greek scientist Aristotle first noticed what we now call M41 back in 325 B.C. He described it simply as a "cloudy spot." Sometime before 1654, Giovanni Hodierna was the first astronomer to catalog it, while Messier independently added M41 to his listing on January 16, 1765.

M41 packs about a hundred stars into an area some 25 light-years across. Several cluster stars are either orange or red giants, including that reddish star visible near the cluster's center. That star turns out to be a spectral-type K giant about 700 times more luminous than the Sun.

## M79: Globular Cluster in Lepus

DISTANCE FROM EARTH: 42,100 light-years

FINDING FACTOR: ***

"WOW!" FACTOR:    Binoculars: *        Small telescopes (3″ to 5″): *
                 Medium telescopes (6″ to 8″): **

### Where to Look

First, locate the dim constellation Lepus the Hare, directly south of Orion. Four stars form a bowl-like trapezoid, while other faint stars curve back toward the northeast to create a sort of backward handle. Extend a line from Arneb through Nihal, the constellation's two brightest stars, and continue an equal distance to the south. You should spot a 5th-magnitude star there, a handy reference. Center on that star, then look in your telescope just to its east for M79, separated by only about half a degree.

### What You'll See

**Through Binoculars.** M79 should be visible through 7 × 50 and 10 × 50 binoculars as a dim, fuzzy patch of light next to that 5th-magnitude star. Although spotting any of its individual stars is an impossible task in anything less than moderately large amateur instruments, binoculars should display the cluster's brighter central concentration.

**Through a Telescope.** The only bright globular cluster in the winter sky that is visible from the Northern Hemisphere, M79 certainly improves with aperture and magnification. The best view will probably be enjoyed using between 100× and 150×. Six-inch (15-cm) and smaller telescopes will show a mottled, grainy texture to M79, as if the cluster is right on the verge of resolution but not quite there. An 8-inch (20-cm) instrument, however, drives it home by revealing a few very faint cluster stars around the fringe (Figure 9.17). Still larger telescopes turn M79 into a splendid sight, with many stars scattered across its disk.

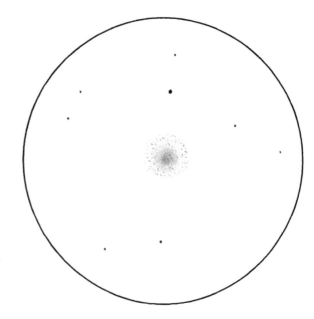

**Figure 9.17** *M79, winter's lone Messier globular cluster, rides low in the southern sky in Lepus. This drawing shows the view through the author's 8-inch reflector at 120×. South is up.*

## M50: Open Cluster in Monoceros

DISTANCE FROM EARTH: 3,000 light-years
FINDING FACTOR:  **
"WOW!" FACTOR:    Binoculars: **        Small telescopes (3″ to 5″): **
                 Medium telescopes (6″ to 8″): **

### Where to Look

Once again, Sirius serves as the starting point for our next quest. As shown on Winter Sky Window 5, Sirius forms a triangle with two fainter stars to its east and northeast, labeled on the map as Mullphen and Theta Canis Majoris. Draw a line from Sirius to Theta and hop an equal distance farther to the northeast. M50 should be bright enough to be visible through most finder-scopes as a small patch of celestial lint. If you can't find Theta, aim your tele-scope about a third of the way from Sirius to the bright star Procyon in Canis Minor, which should put you in M50's vicinity.

### What You'll See

**Through Binoculars.** Although it appears to be located out in the middle of nowhere, M50 is surprisingly easy to find in 7× binoculars, and may even reach naked eye visibility on dark evenings. Only half a dozen or so of its many stars are bright enough for binocular viewing, however, the remainder blending into a pleasant background glow. Look for a reddish star south of the cluster's center.

**Through a Telescope.** Even though its home constellation is faint, M50 is bright enough to be easily visible in just about any telescope. Use low power for the best view. A 3-inch (7.5-cm) telescope will show around twenty-five stars within the cluster, while a 6-inch (15-cm) will multiply that number by at least a factor of two. The brighter cluster stars seem to fall into curved lines and arcs that some see as distinctively heart-shaped or possibly like an arrowhead oriented east-west. A small triangle of stars also marks the cluster's center. While most of the stars in M50 appear white or blue-white, a lone red star highlights the cluster a little south of center. If you have trouble making out its color, try defocusing the image slightly.

## M47: Open Cluster in Puppis

DISTANCE FROM EARTH: 1,600 light-years
FINDING FACTOR: **
"WOW!" FACTOR:    Binoculars: ***    Small telescopes (3″ to 5″): ****
Medium telescopes (6″ to 8″): ****

### Where to Look

Finding M47 is a matter of following the triangles, three to be exact. Aim your finderscope at Sirius, brightest star in the night sky. As shown on Sky Window 5, Sirius forms a narrow triangle with the fainter stars Iota Canis Majoris and Mullphen to its east. Follow the Sirius-Mullphen side of the triangle to the east, to a second, smaller narrow triangle, then farther eastward still to a third. While all three triangles have similar profiles, the third, easternmost triangle should also show a few more interior stars than the other two. That's because the third triangle also surrounds both M46 (described below) and M47.

### What You'll See

**Through Binoculars.** M47 is easy to spot through most binoculars as a hazy blotch peppered with several pinpoints of light. The cluster itself is quite pretty, but when combined with the starry surroundings, the overall impact is striking! Take your time and enjoy the view.

**Through a Telescope.** M47 is a wonderful open cluster to view with low power. Thirty blue-white stars ranging in brightness from 6th to 12th magnitude belong to the group and are bright enough to be resolved through a 6-inch (15-cm) telescope. The beauty of the scene is further enhanced by surroundings that overflow with stars.

Be sure to take note of the star that marks the western corner of the triangle surrounding M47. That's the reddish orange variable star KQ Puppis, a solitary ruby buried among sapphires.

## M46: Open Cluster in Puppis

DISTANCE FROM EARTH: 5,400 light-years
FINDING FACTOR: **
"WOW!" FACTOR:    Binoculars: **    Small telescopes (3″ to 5″): **
Medium telescopes (6″ to 8″): ***

### Where to Look

Find M47 using the directions above, then look for M46 just to the east. Reorient yourself by looking through the finderscope for the slender triangle that surrounded M47. M46 is inside the same triangle but about a degree (two Full Moons) to the east. You might be able to spot it through your finderscope and even squeeze both into the same low-power eyepiece field.

### What You'll See

**Through Binoculars.** Unlike M47, which has several stars within the grasp of most binoculars, neighboring M46 is a rich congregation of very faint stars. As a result, most binoculars will show only a hazy glow, although giant glasses 70 mm and larger in aperture may reveal a few very faint specks within the wispy cloud.

**Through a Telescope.** Here's a good example of just how different open clusters can be. The previous object, M47, is a coarse group made up of many bright stars, while M46 is much denser, literally bulging at the seams with stars. But because the stars are also quite faint, their light blends into a hazy cloud through smaller telescopes. A good quality 4-inch (10-cm) telescope should be able to reveal a few faint grains of starlight poking through the dim glow, while larger telescopes resolve more and more individual cluster members.

M46 is also unique because of a celestial interloper that lies in front of it. If you have a 6-inch (15-cm) or larger telescope, look for a tiny, gray disk among the stars in the northern part of the cluster. That disk is NGC 2438, a planetary nebula that shines faintly at 10th magnitude. It's a tough catch because of its small size, so be sure to use between 150× and 200× to pick it out.

### M46

The main reason that the stars in M46 look so much fainter than those in M47 is the difference in distance. M46 is over three times farther away than its apparent neighbor, M47. More than five hundred stars scattered across 30 light-years make up M46. The brightest, shining around 9th magnitude, are hot, spectral-type A blue-white giants, each a powerhouse over 100 times more luminous than our Sun.

Although planetary nebula NGC 2438 looks like it belongs to M46, in reality it is much closer, "only" about 2,900 light-years from Earth. One reason we know that is by comparing the spectra of the planetary with that of some of the stars in M46. These show that both M46 and NGC 2438 are moving away from us as we pursue our independent paths around the Milky Way, but at different speeds. Were the planetary and cluster physically connected, they would be moving through space at the same speed. Another even more obvious reason is that stars in open clusters are usually quite young, in contrast to planetary nebulae, which form from stars that are very old.

M46 was discovered by Charles Messier on February 19, 1771, only three days after he had published the first edition of his catalog, which covered M's 1 through 45.

### M93: Open Cluster in Puppis

DISTANCE FROM EARTH: 3,600 light-years

FINDING FACTOR: ***

"WOW!" FACTOR:    Binoculars: **      Small telescopes (3″ to 5″): ***
                 Medium telescopes (6″ to 8″): ***

#### Where to Look

Although M93 is bright enough to be seen through binoculars, its southerly position as well as remote location in the sky can make finding it a challenge. Begin again at Sirius, then trace out the Dog's stick body toward the southern horizon. Follow its back to the star Wezen, and its two legs, marked by the stars Adhara and Aludra. Aim your finder at the latter, then follow a curve of four stars that hooks toward the northeast. The fourth star in the curve also marks one of four corners in a crooked rectangle. Center the brightest star in the rectangle, golden Xi Puppis, in your finderscope, then look just toward its northwest. M93 should be visible as a dim smudge of light just beyond a small triangle of faint stars.

#### What You'll See

**Through Binoculars.** M93 is a pretty sight in nearly all binoculars. Through 50-mm binoculars, you might spot a half dozen or so stars mixed throughout a dim glow that, to my eyes, appears triangular.

**Through a Telescope.** Appearing smaller than either M46 or M47, M93 nonetheless puts on a fine show through backyard telescopes. A 4-inch (10-cm) instrument should reveal between twenty-five and thirty faint stars, with many of them strung together in curves and arcs against the soft glow of still fainter suns. About half a dozen of the brighter members form a line that zigzags close to the cluster's center, while the remaining suns fill in to give the group a triangular shape overall. Use a low- to medium-power eyepiece for the best view.

## Winter Sky Window 6

Bridging the gap between Canis Major and the head of Hydra, Winter Sky Window 6 (Figure 9.18) looks nearly barren to the eye alone. Both constellations prove handy for finding the one new Messier object here, M48.

### M48: Open Cluster in Hydra

DISTANCE FROM EARTH: 1,500 light-years

FINDING FACTOR: **

"WOW!" FACTOR:    Binoculars: **      Small telescopes (3″ to 5″): ***
                 Medium telescopes (6″ to 8″): ***

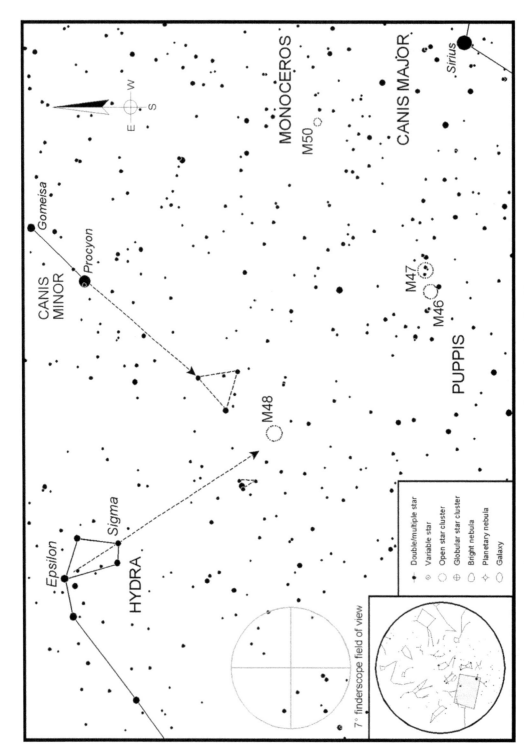

**Figure 9.18** *Winter Sky Window 6 bridges the gap between the late winter and early spring skies.*

### Where to Look

Even though M48 is set in an empty part of the late winter sky, it is surprisingly easy to find. First, locate the bright star Procyon and its tiny constellation Canis Minor, the Small Dog. Extend a line from the star Gomeisa through Procyon toward the southeast. Take about two hops until you see an equilateral triangle of stars. Continue on the same course. M48 lies just beyond the triangle and should be visible through your finderscope as a dim patch of starlight.

M48 can also be found by locating the dim head of springtime's Hydra the Serpent. Use the stars in the head like an arrow, as shown in Winter Sky Window 6. Draw a line from Epsilon through Sigma and continue it toward the southwest. Viewing through your finderscope, look for a fairly bright star surrounded by a small triangle of fainter suns. That pattern lies a little to the east of M48. If you put the triangle in the center of your finderscope, M48 should just squeeze into the southwestern edge.

### What You'll See

**Through Binoculars.** M48 is a very pretty cluster through binoculars. Look for a dim haze of unresolved starlight surrounding a few faint points of light. The brightest member of M48 shines at 8th magnitude and so should be bright enough to be spotted through 50-mm binoculars. From dark skies, you might also glimpse three stars in a triangular pattern at its center. The remaining cluster stars blend into a homogeneous glow that stands out well against the surrounding sky.

**Through a Telescope.** Since M48 covers about half a degree of sky, a low-power eyepiece is the best choice for enjoying the view. Once the cluster is centered, you should see M48 as a bright, concentrated collection of more than fifty stars, many of which appear set in lines and arcs. A tight grouping of eight suns resembling a capital A highlights the center of the cluster, while outliers seem to give the cluster an overall triangular shape, looking something like an arrowhead that is aimed toward the east.

### *M48*

Charles Messier discovered M48 in 1771, but he made a mistake when noting its position in the sky, erroneously reporting its location as 4° farther north than it actually is. This bungle caused M48 to be listed as "lost" until 1959. That year, T. F. Morris of the Royal Astronomical Society of Canada, while examining Messier's records, suggested that the only open cluster in the area that agreed with Messier's description was NGC 2548. Today, they are considered one and the same.

This cluster of eighty stars measures about 23 light-years in diameter. Most of the stars are type B blue giants, although three yellow giants have also been found. Their average age is believed to be about 300 million years.

# Epilogue

## Afterthoughts

By the time you have reached this point in the book, you will have had a chance to go out and try your luck with some of the many objects discussed earlier, whether it was the Moon, planets, Sun, or distant deep-sky objects. I hope you take the time to observe each, taking notes along the way. Once you have successfully found all of the objects in this book, your skills as an observer should be sharp. After all, not every amateur astronomer takes the time to find all of the Messier objects. By following this book to completion, you have not only accomplished that feat, but you have also seen several other deep-sky objects as well, to say nothing of the Sun, the Moon, and the planets. Congratulations!

Often astronomy clubs have several observing programs designed to encourage members to view a set of objects with their own telescopes, with certificates of accomplishment awarded to those who complete each list. It's a nice way to acknowledge a person's achievement, and let's face it, everyone likes an award. With that in mind, I would like to extend the following offer. If and when you view every deep-sky object in this book, I would like to hear from you. Send an e-mail to phil@philharrington.net and tell me about your accomplishment. Include the following information: your name, town, state/ country, equipment used, and how long it took you to complete all four seasonal lists (that is, those in chapters 6 through 9). By return e-mail, I will send you a Star Watcher certificate suitable for color printing and framing that is numbered and signed attesting to your achievement. If you prefer, you may contact me by old-fashioned snail mail by sending the requested information to me in care of the publisher, John Wiley & Sons (address in the front of the book). If you include a self-addressed, stamped return envelope, I'll be happy to send you the same certificate. Make sure that the envelope is large enough to handle an 8.5″-by-11″ sheet of paper. I will also publish on my Web site (www.philharrington.net) a list of everyone who completes the requirements for a Star Watcher certificate, unless you would prefer otherwise. Just tell me in your e-mail or letter. I will only list your name, your town, and the date you finished the list, not your contact information (which also will not be sold or given to anyone, commercial or otherwise; in fact, I will delete it as soon as I send out your certificate). Just to cover myself and to make sure that my heirs don't have to worry about it, I must add that this offer can be retracted at any time without prior notice.

Beyond this, once you complete all of the objects this book has to offer, you may be left asking, Okay, what next? That's a very good question. The answer depends on you. What did you enjoy doing most? Observing solar system objects? If so, join the Association of Lunar and Planetary Observers (ALPO). Prefer deep-sky objects? Then pick up where you left off with the *New General Catalog* (NGC), which lists more than 7,800 targets. That should keep you busy for a while.

For those who prefer a more structured program, both the Astronomical League and the Royal Astronomical Society of Canada (see contact information in Bibliography and References) have a series of observing clubs, with very nice certificates and pins given to their members who complete the lists. Some of the observing programs hosted by the AL and RASC include the Asteroid Observing Club, Binocular Messier Club, Double Star Club, Herschel 400 Club, Lunar Club, Messier Club, Planetary Observers Club, and Urban Observing Club. Many who polish off all of the Messier objects enjoy the challenge of the Herschel 400, so you might consider that as your next great observing assignment.

As you see, there are many more celestial sights for you to find and do. This book was only the beginning. If it's clear out tonight and you have already observed everything that we have discussed here, put this book down. Then get out the telescope and charts, and start anew. The rest of the universe is waiting for you.

# Star Watcher Award

Presented to

_____

for observing all of the deep-sky objects, including all of the Messier objects, listed in the book *Star Watch*.

Awarded on this ____ day of _____

Certificate number  001

Philip S. Harrington
Author, *Star Watch*

*Star Watcher certificate.*

# Appendix A
# The Constellations

| Constellation | Abbreviation | Genitive | Season[1] | Meaning |
|---|---|---|---|---|
| Andromeda | And | Andromedae | Autumn | Princess |
| Antlia | Ant | Antliae | (S) | Air Pump |
| Apus | Aps | Apodis | (S) | Bird of Paradise |
| Aquarius | Aqr | Aquarii | Autumn | The Water Bearer |
| Aquila | Aql | Aquilae | Summer | Eagle |
| Ara | Ara | Arae | Summer | Altar |
| Aries | Ari | Arietis | Autumn | Ram |
| Auriga | Aur | Aurigae | Winter | Charioteer |
| Boötes | Boo | Boötis | Spring | Herdsman |
| Caelum | Cae | Caeli | (S) | Chisel |
| Camelopardalis | Cam | Camelopardalis | Winter (C) | Giraffe |
| Cancer | Cnc | Cancri | Spring | Crab |
| Canes Venatici | Cvn | Canum Venaticorum | Spring | Hunting Dogs |
| Canis Major | CMa | Canis Majoris | Winter | Big Dog |
| Canis Minor | CMi | Canis Minoris | Winter | Little Dog |
| Capricornus | Cap | Capricorni | Autumn | Sea-Goat |
| Carina | Car | Carinae | (S) | Keel (of the mythical ship *Argo*) |
| Cassiopeia | Cas | Cassiopeiae | Autumn (C) | Queen |
| Centaurus | Cen | Centauri | (S) | Centaur |
| Cepheus | Cep | Cephei | Autumn (C) | King |
| Cetus | Cet | Ceti | Autumn | Whale or Sea Monster |
| Chamaeleon | Cha | Chamaeleontis | (S) | Chameleon |
| Circinus | Cir | Circini | (S) | Compasses |
| Columba | Col | Columbae | Winter | Dove |
| Coma Berenices | Com | Comae Berenices | Spring | Queen Berenice's Hair |
| Corona Australis | CrA | Coronae Australis | (S) | Southern Crown |
| Corona Borealis | CrB | Coronae Borealis | Spring | Northern Crown |
| Corvus | Crv | Corvi | Spring | Crow |
| Crater | Crt | Crateris | Spring | Cup |
| Crux | Cru | Crucis | (S) | Southern Cross |
| Cygnus | Cyg | Cygni | Summer | Swan |
| Delphinus | Del | Delphini | Summer | Dolphin |
| Dorado | Dor | Doradus | (S) | Swordfish |
| Draco | Dra | Draconis | Summer (C) | Dragon |
| Equuleus | Equ | Equulei | Spring | Little Horse |
| Eridanus | Eri | Eridani | Winter | River |
| Fornax | For | Fornacis | Autumn | Furnace |
| Gemini | Gem | Geminorum | Winter | Twins |
| Hercules | Her | Herculis | Summer | Giant |

*(continued)*

| Constellation | Abbreviation | Genitive | Season[1] | Meaning |
|---|---|---|---|---|
| Horologium | Hor | Horologii | (S) | Clock |
| Hydra | Hya | Hydrae | Spring | Water Snake (male) |
| Hydrus | Hyi | Hydri | (S) | Water Snake (female) |
| Indus | Ind | Indi | (S) | Indian |
| Lacerta | Lac | Lacertae | Autumn | Lizard |
| Leo | Leo | Leonis | Spring | Lion |
| Leo Minor | LMi | Leonis Minoris | Spring | Little Lion |
| Lepus | Lep | Leporis | Winter | Hare |
| Libra | Lib | Librae | Summer | Scales of Justice |
| Lupus | Lup | Lupi | (S) | Wolf |
| Lynx | Lyn | Lyncis | Winter | Lynx |
| Lyra | Lyr | Lyrae | Summer | Lyre |
| Mensa | Men | Mensae | (S) | Table |
| Microscopium | Mic | Microscopii | (S) | Microscope |
| Monoceros | Mon | Monocerotis | Winter | Unicorn |
| Musca | Mus | Muscae | (S) | Fly |
| Norma | Nor | Normae | (S) | Square |
| Octans | Oct | Octantis | (S) | Octant |
| Ophiuchus | Oph | Ophiuchi | Summer | Serpent Bearer |
| Orion | Ori | Orionis | Winter | Hunter |
| Pavo | Pav | Pavonis | (S) | Peacock |
| Pegasus | Peg | Pegasi | Autumn | Flying Horse |
| Perseus | Per | Persei | Autumn | Warrior |
| Phoenix | Phe | Phoenicis | (S) | Phoenix |
| Pictor | Pic | Pictoris | (S) | Painter |
| Pisces | Psc | Piscium | Autumn | Fishes |
| Piscis Austrinus | PsA | Piscis Austrini | Autumn | Southern Fish |
| Puppis | Pup | Puppis | Winter | Stern (of the mythical ship *Argo*) |
| Pyxis | Pyx | Pyxidis | Spring | Compass |
| Reticulum | Ret | Reticuli | (S) | Reticle |
| Sagitta | Sge | Sagittae | Summer | Arrow |
| Sagittarius | Sgr | Sagittarii | Summer | Archer |
| Scorpius | Sco | Scorpii | Summer | Scorpion |
| Sculptor | Scl | Sculptoris | Autumn | Sculptor |
| Scutum | Sct | Scuti | Summer | Shield |
| Serpens | Ser | Serpentis | Summer | Serpent |
| Sextans | Sex | Sextantis | Spring | Sextant |
| Taurus | Tau | Tauri | Winter | Bull |
| Telescopium | Tel | Telescopii | (S) | Telescope |
| Triangulum | Tri | Trianguli | Autumn | Triangle |
| Triangulum Australe | TrA | Trianguli Australis | (S) | Southern Triangle |
| Tucana | Tuc | Tucanae | (S) | Toucan |
| Ursa Major | UMa | Ursae Majoris | Spring (C) | Great Bear |
| Ursa Minor | UMi | Ursae Minoris | Summer (C) | Little Bear |
| Vela | Vel | Velorum | (S) | Sails (of the mythical ship *Argo*) |
| Virgo | Vir | Virginis | Spring | Maiden |
| Volans | Vol | Volantis | (S) | Flying Fish |
| Vulpecula | Vul | Vulpeculae | Summer | Fox |

1. *Season based on midnorthern latitudes unless otherwise noted. Circumpolar constellations are listed with the season when they are highest in the sky but are denoted with (C) to note their year-round visibility. Those constellations located far below the Celestial Equator, and hence only well seen to the south of Earth's equator, are listed with their Southern Hemisphere season of visibility as well as (S) to show their far-southern position in the sky.*

# Appendix B
# Planets: 2003–2015

The listing below gives the midmonth positions of the naked-eye planets (excluding Mercury, since it moves so rapidly in the sky) through the year 2015. Constellation abbreviations are defined in Appendix A. Although the Sun follows the ecliptic precisely, the planets can stray beyond the boundaries of the twelve traditional Zodiacal constellations. They can also pass into portions of Ophiuchus (Oph), Cetus (Cet), Sextans (Sex), Orion (Ori), and Hydra (Hya).

**2003**

|         | Jan | Feb | Mar | Apr | May | Jun | Jul | Aug | Sep | Oct | Nov | Dec |
|---------|-----|-----|-----|-----|-----|-----|-----|-----|-----|-----|-----|-----|
| Venus   | Oph | Sgr | Cap | Psc | Ari | Tau | Gem | Leo | Vir | Lib | Oph | Sgr |
| Mars    | Lib | Oph | Sgr | Sgr | Cap | Aqr | Aqr | Aqr | Aqr | Aqr | Aqr | Psc |
| Jupiter | Cnc | Cnc | Cnc | Cnc | Cnc | Cnc | Leo | Leo | Leo | Leo | Leo | Leo |
| Saturn  | Tau | Tau | Tau | Tau | Ori | Gem | Gem | Gem | Gem | Gem | Gem | Gem |

**2004**

|         | Jan | Feb | Mar | Apr | May | Jun | Jul | Aug | Sep | Oct | Nov | Dec |
|---------|-----|-----|-----|-----|-----|-----|-----|-----|-----|-----|-----|-----|
| Venus   | Aqr | Psc | Ari | Tau | Tau | Tau | Tau | Gem | Cnc | Leo | Vir | Lib |
| Mars    | Psc | Ari | Tau | Tau | Gem | Gem | Cnc | Leo | Leo | Vir | Vir | Lib |
| Jupiter | Leo | Leo | Leo | Leo | Leo | Leo | Leo | Leo | Vir | Vir | Vir | Vir |
| Saturn  | Gem | Gem | Gem | Gem | Gem | Gem | Gem | Gem | Gem | Gem | Gem | Gem |

**2005**

|         | Jan | Feb | Mar | Apr | May | Jun | Jul | Aug | Sep | Oct | Nov | Dec |
|---------|-----|-----|-----|-----|-----|-----|-----|-----|-----|-----|-----|-----|
| Venus   | Sgr | Cap | Psc | Ari | Tau | Gem | Leo | Vir | Vir | Oph | Sgr | Cap |
| Mars    | Oph | Sgr | Sgr | Cap | Aqr | Psc | Psc | Ari | Ari | Ari | Ari | Ari |
| Jupiter | Vir | Vir | Vir | Vir | Vir | Vir | Vir | Vir | Vir | Vir | Vir | Lib |
| Saturn  | Gem | Gem | Gem | Gem | Gem | Gem | Cnc | Cnc | Cnc | Cnc | Cnc | Cnc |

**2006**

|         | Jan | Feb | Mar | Apr | May | Jun | Jul | Aug | Sep | Oct | Nov | Dec |
|---------|-----|-----|-----|-----|-----|-----|-----|-----|-----|-----|-----|-----|
| Venus   | Sgr | Sgr | Cap | Aqr | Psc | Ari | Tau | Cnc | Leo | Vir | Lib | Sgr |
| Mars    | Ari | Tau | Tau | Gem | Gem | Cnc | Leo | Leo | Vir | Vir | Lib | Sco |
| Jupiter | Lib | Lib | Lib | Lib | Lib | Lib | Lib | Lib | Lib | Lib | Lib | Sco |
| Saturn  | Cnc | Cnc | Cnc | Cnc | Cnc | Cnc | Cnc | Cnc | Leo | Leo | Leo | Leo |

**2007**

|         | Jan | Feb | Mar | Apr | May | Jun | Jul | Aug | Sep | Oct | Nov | Dec |
|---------|-----|-----|-----|-----|-----|-----|-----|-----|-----|-----|-----|-----|
| Venus   | Cap | Psc | Ari | Tau | Gem | Cnc | Leo | Sex | Cnc | Leo | Vir | Lib |
| Mars    | Sgr | Sgr | Cap | Aqr | Psc | Psc | Ari | Tau | Tau | Gem | Gem | Gem |
| Jupiter | Oph | Oph | Oph | Oph | Oph | Oph | Oph | Oph | Oph | Oph | Oph | Sgr |
| Saturn  | Leo | Leo | Leo | Leo | Leo | Leo | Leo | Leo | Leo | Leo | Leo | Leo |

*(continued)*

**2008**

|         | Jan | Feb | Mar | Apr | May | Jun | Jul | Aug | Sep | Oct | Nov | Dec |
|---------|-----|-----|-----|-----|-----|-----|-----|-----|-----|-----|-----|-----|
| Venus   | Sgr | Cap | Aqr | Psc | Tau | Gem | Cnc | Vir | Vir | Sco | Sgr | Cap |
| Mars    | Tau | Tau | Gem | Gem | Cnc | Leo | Leo | Vir | Vir | Vir | Sco | Oph |
| Jupiter | Sgr | Sgr | Sgr | Sgr | Sgr | Sgr | Sgr | Sgr | Sgr | Sgr | Sgr | Sgr |
| Saturn  | Leo | Leo | Leo | Leo | Leo | Leo | Leo | Leo | Leo | Leo | Leo | Leo |

**2009**

|         | Jan | Feb | Mar | Apr | May | Jun | Jul | Aug | Sep | Oct | Nov | Dec |
|---------|-----|-----|-----|-----|-----|-----|-----|-----|-----|-----|-----|-----|
| Venus   | Aqr | Psc | Mar | Psc | Psc | Ari | Tau | Gem | Leo | Vir | Lib | Oph |
| Mars    | Sgr | Cap | Aqr | Psc | Psc | Ari | Tau | Tau | Gem | Cnc | Cnc | Leo |
| Jupiter | Cap | Cap | Cap | Cap | Cap | Cap | Cap | Cap | Cap | Cap | Cap | Cap |
| Saturn  | Leo | Leo | Leo | Leo | Leo | Leo | Leo | Leo | Vir | Vir | Vir | Vir |

**2010**

|         | Jan | Feb | Mar | Apr | May | Jun | Jul | Aug | Sep | Oct | Nov | Dec |
|---------|-----|-----|-----|-----|-----|-----|-----|-----|-----|-----|-----|-----|
| Venus   | Sgr | Aqr | Psc | Ari | Tau | Cnc | Leo | Vir | Vir | Lib | Vir | Lib |
| Mars    | Cnc | Cnc | Cnc | Cnc | Leo | Leo | Leo | Vir | Vir | Lib | Oph | Sgr |
| Jupiter | Aqr | Aqr | Aqr | Aqr | Psc | Psc | Psc | Psc | Psc | Psc | Aqr | Aqr |
| Saturn  | Vir | Vir | Vir | Vir | Vir | Vir | Vir | Vir | Vir | Vir | Vir | Vir |

**2011**

|         | Jan | Feb | Mar | Apr | May | Jun | Jul | Aug | Sep | Oct | Nov | Dec |
|---------|-----|-----|-----|-----|-----|-----|-----|-----|-----|-----|-----|-----|
| Venus   | Oph | Sgr | Cap | Aqr | Psc | Tau | Gem | Leo | Vir | Lib | Oph | Sgr |
| Mars    | Cap | Cap | Aqr | Psc | Ari | Tau | Tau | Gem | Cnc | Cnc | Leo | Leo |
| Jupiter | Psc | Psc | Psc | Psc | Psc | Ari | Ari | Ari | Ari | Ari | Ari | Ari |
| Saturn  | Vir | Vir | Vir | Vir | Vir | Vir | Vir | Vir | Vir | Vir | Vir | Vir |

**2012**

|         | Jan | Feb | Mar | Apr | May | Jun | Jul | Aug | Sep | Oct | Nov | Dec |
|---------|-----|-----|-----|-----|-----|-----|-----|-----|-----|-----|-----|-----|
| Venus   | Aqr | Psc | Ari | Tau | Tau | Tau | Tau | Gem | Cnc | Leo | Vir | Lib |
| Mars    | Leo | Leo | Leo | Leo | Leo | Leo | Vir | Vir | Lib | Sco | Sgr | Sgr |
| Jupiter | Ari | Ari | Ari | Ari | Ari | Tau | Tau | Tau | Tau | Tau | Tau | Tau |
| Saturn  | Vir | Vir | Vir | Vir | Vir | Vir | Vir | Vir | Vir | Vir | Vir | Lib |

**2013**

|         | Jan | Feb | Mar | Apr | May | Jun | Jul | Aug | Sep | Oct | Nov | Dec |
|---------|-----|-----|-----|-----|-----|-----|-----|-----|-----|-----|-----|-----|
| Venus   | Sgr | Cap | Aqr | Ari | Tau | Gem | Leo | Vir | Vir | Oph | Sgr | Sgr |
| Mars    | Cap | Aqr | Psc | Psc | Ari | Tau | Gem | Gem | Cnc | Leo | Leo | Vir |
| Jupiter | Tau | Tau | Tau | Tau | Tau | Tau | Gem | Gem | Gem | Gem | Gem | Gem |
| Saturn  | Lib | Lib | Lib | Lib | Vir | Vir | Vir | Vir | Lib | Lib | Lib | Lib |

**2014**

|         | Jan | Feb | Mar | Apr | May | Jun | Jul | Aug | Sep | Oct | Nov | Dec |
|---------|-----|-----|-----|-----|-----|-----|-----|-----|-----|-----|-----|-----|
| Venus   | Sgr | Sgr | Cap | Aqr | Psc | Ari | Tau | Cnc | Leo | Vir | Lib | Sgr |
| Mars    | Vir | Vir | Vir | Vir | Vir | Vir | Vir | Lib | Sco | Oph | Sgr | Cap |
| Jupiter | Gem | Gem | Gem | Gem | Gem | Gem | Cnc | Cnc | Cnc | Leo | Leo | Leo |
| Saturn  | Lib | Lib | Lib | Lib | Lib | Lib | Lib | Lib | Lib | Lib | Lib | Lib |

**2015**

|         | Jan | Feb | Mar | Apr | May | Jun | Jul | Aug | Sep | Oct | Nov | Dec |
|---------|-----|-----|-----|-----|-----|-----|-----|-----|-----|-----|-----|-----|
| Venus   | Cap | Aqr | Psc | Tau | Gem | Cnc | Leo | Leo | Cnc | Leo | Vir | Lib |
| Mars    | Aqr | Psc | Psc | Ari | Tau | Tau | Cnc | Cnc | Leo | Leo | Vir | Vir |
| Jupiter | Leo | Cnc | Cnc | Cnc | Cnc | Leo | Leo | Leo | Leo | Leo | Leo | Leo |
| Saturn  | Sco | Sco | Sco | Sco | Lib | Lib | Lib | Lib | Lib | Lib | Sco | Oph |

# Appendix C

# The Messier Catalog Plus

All of the deep-sky objects that are described in chapters 5 through 9 are included in this data listing. Columns from left to right include the Sky Window listing each is found on, the object's name or catalog number, its constellation (see Appendix A for definitions), right ascension and declination coordinates, magnitude, apparent size (in lieu of size, all binary and multiple stars list their separation, while variable stars list their periods), and comments, such as an object's nickname. Data are sorted by season, Sky Window, and finally by increasing right ascension within each window.

| Sky Window | Object | Constellation¹ | Type² | Right Ascension | | Declination | | Magnitude | Size³ | Comment |
|---|---|---|---|---|---|---|---|---|---|---|
| | | | | hh | mm.m | ° | ' | | | |
| **Spring** | | | | | | | | | | |
| Spring 1 | M44 (NGC 2632) | Cnc | OC | 08 | 40.1 | +19 | 59 | 3.1 | 95' | Beehive or Praesepe |
| Spring 1 | Iota | Cnc | *** | 08 | 46.7 | +28 | 46 | 4.0, 6.6 | 30" | |
| Spring 1 | M67 (NGC 2682) | Cnc | OC | 08 | 50.4 | +11 | 49 | 6.1 | 30' | |
| Spring 2 | NGC 2976 | UMa | Gx | 09 | 47.3 | +67 | 55 | 10.2 | 5' | |
| Spring 2 | M81 (NGC 3031) | UMa | Gx | 09 | 55.6 | +69 | 04 | 7.0 | 26' × 14' | Bode's Nebula |
| Spring 2 | M82 (NGC 3034) | UMa | Gx | 09 | 55.8 | +69 | 41 | 8.4 | 11' × 5' | Cigar Galaxy |
| Spring 2 | NGC 3077 | UMa | Gx | 10 | 03.3 | +68 | 44 | 9.9 | 5' | |
| Spring 2 | M108 (NGC 3556) | UMa | Gx | 11 | 11.5 | +55 | 40 | 10.1 | 8' × 3' | |
| Spring 2 | M97 (NGC 3587) | UMa | PN | 11 | 14.8 | +55 | 01 | 11.2 | 3' | |
| Spring 2 | M109 (NGC 3992) | UMa | Gx | 11 | 57.6 | +53 | 23 | 9.8 | 8' × 5' | |
| Spring 2 | M40 | UMa | ** | 12 | 22.4 | +58 | 05 | 9.0, 9.3 | 50" | Winnecki 4 |
| Spring 3 | M106 (NGC 4258) | CVn | Gx | 12 | 19.0 | +47 | 18 | 8.3 | 18' × 8' | |
| Spring 3 | M94 (NGC 4736) | CVn | Gx | 12 | 50.9 | +41 | 07 | 8.2 | 11' × 9' | |
| Spring 3 | M63 (NGC 5055) | CVn | Gx | 13 | 15.8 | +42 | 02 | 8.6 | 12' × 7' | Sunflower Galaxy |
| Spring 3 | Alcor/Mizar | UMa | ** | 13 | 23.9 | +54 | 56 | 2.3, 4.0 | 14" | |

(continued)

| Sky Window | Object | Constellation[1] | Type[2] | Right Ascension hh | mm.m | Declination ° | ' | Magnitude | Size[3] | Comment |
|---|---|---|---|---|---|---|---|---|---|---|
| Spring 3 | M51 (NGC 5194) | CVn | Gx | 13 | 29.9 | +47 | 12 | 8.4 | 11' × 8' | Whirlpool Galaxy |
| Spring 3 | NGC 5195 | CVn | Gx | 13 | 30.0 | +47 | 16 | 9.6 | 5.4' | M51 companion |
| Spring 3 | M101 (NGC 5457) | UMa | Gx | 14 | 03.2 | +54 | 21 | 7.7 | 27' × 26' | Pinwheel Galaxy |
| Spring 3 | M102 (NGC 5866) | Dra | Gx | 15 | 06.5 | +55 | 46 | 9.9 | 5' × 2' | Spindle Galaxy; see note 4 |
| Spring 4 | NGC 2903 | Leo | Gx | 09 | 32.2 | +21 | 30 | 8.9 | 13' | |
| Spring 4 | M95 (NGC 3351) | Leo | Gx | 10 | 44.0 | +11 | 42 | 9.7 | 7' × 5' | |
| Spring 4 | M96 (NGC 3368) | Leo | Gx | 10 | 46.8 | +11 | 49 | 9.2 | 7' × 5' | |
| Spring 4 | M105 (NGC 3379) | Leo | Gx | 10 | 47.8 | +12 | 35 | 9.3 | 5' × 4' | |
| Spring 4 | M65 (NGC 3623) | Leo | Gx | 11 | 18.9 | +13 | 05 | 9.3 | 10' × 3' | |
| Spring 4 | M66 (NGC 3627) | Leo | Gx | 11 | 20.2 | +12 | 59 | 9.0 | 9' × 4' | |
| Spring 4 | NGC 3628 | Leo | Gx | 11 | 20.3 | +13 | 36 | 9.5 | 14' | |
| Spring 5 | M98 (NGC 4192) | Com | Gx | 12 | 13.8 | +14 | 54 | 10.1 | 10' × 3' | |
| Spring 5 | M99 (NGC 4254) | Com | Gx | 12 | 18.8 | +14 | 25 | 9.8 | 5' × 4' | The Pinwheel |
| Spring 5 | M61 (NGC 4303) | Vir | Gx | 12 | 21.9 | +04 | 28 | 9.7 | 6' × 5' | |
| Spring 5 | M100 (NGC 4321) | Com | Gx | 12 | 22.9 | +15 | 49 | 9.4 | 7' × 6' | |
| Spring 5 | M84 (NGC 4374) | Vir | Gx | 12 | 25.1 | +12 | 53 | 9.3 | 5' × 4' | |
| Spring 5 | M85 (NGC 4382) | Com | Gx | 12 | 25.4 | +18 | 11 | 9.2 | 7' × 5' | |
| Spring 5 | NGC 4387 | Vir | Gx | 12 | 25.7 | +12 | 49 | 12.0 | 2' | |
| Spring 5 | NGC 4388 | Vir | Gx | 12 | 25.8 | +12 | 40 | 11.0 | 6' | |
| Spring 5 | M86 (NGC 4406) | Vir | Gx | 12 | 26.2 | +12 | 57 | 9.2 | 7' × 6' | |
| Spring 5 | M49 (NGC 4472) | Vir | Gx | 12 | 29.8 | +08 | 00 | 8.4 | 9' × 7' | |
| Spring 5 | M87 (NGC 4486) | Vir | Gx | 12 | 30.8 | +12 | 24 | 8.6 | 7' | Virgo A radio source |
| Spring 5 | M88 (NGC 4501) | Com | Gx | 12 | 32.0 | +14 | 25 | 9.5 | 7' × 4' | |
| Spring 5 | M91 (NGC 4548) | Com | Gx | 12 | 35.4 | +14 | 30 | 10.2 | 4' × 3' | |
| Spring 5 | M89 (NGC 4552) | Vir | Gx | 12 | 35.7 | +12 | 33 | 9.8 | 4' | |
| Spring 5 | M90 (NGC 4569) | Vir | Gx | 12 | 36.8 | +13 | 10 | 9.5 | 9' × 5' | |
| Spring 5 | M58 (NGC 4579) | Vir | Gx | 12 | 37.7 | +11 | 49 | 9.8 | 5' × 4' | |
| Spring 5 | M59 (NGC 4621) | Vir | Gx | 12 | 42.0 | +11 | 39 | 9.8 | 5' × 3' | |
| Spring 5 | M60 (NGC 4649) | Vir | Gx | 12 | 43.7 | +11 | 33 | 8.8 | 7' × 6' | |
| Spring 5 | NGC 4762 | Vir | Gx | 12 | 52.9 | +11 | 14 | 10.3 | 9' | |
| Spring 6 | M68 (NGC 4590) | Hya | GC | 12 | 39.5 | -26 | 45 | 8.2 | 12' | |
| Spring 6 | M104 (NGC 4594) | Vir | Gx | 12 | 40.0 | -11 | 37 | 8.3 | 9' × 4' | Sombrero Galaxy |
| Spring 6 | M83 (NGC 5236) | Hya | Gx | 13 | 37.0 | -29 | 52 | 7.6 | 11' × 10' | Southern Pinwheel Galaxy |
| Spring 7 | Coma Star Cluster | Com | OC | 12 | 25.0 | +26 | | 1.8 | 275' | |
| Spring 7 | M64 (NGC 4826) | Com | Gx | 12 | 56.7 | +21 | 41 | 8.5 | 9' × 5' | Black-Eye Galaxy |
| Spring 7 | M53 (NGC 5024) | Com | GC | 13 | 12.9 | +18 | 10 | 7.7 | 13' | |
| Spring 7 | M3 (NGC 5272) | CVn | GC | 13 | 42.2 | +28 | 23 | 6.4 | 16' | |

| Season | Object | Const | Type | RA h | RA m | Dec ° | Dec ' | Mag | Size | Name |
|---|---|---|---|---|---|---|---|---|---|---|
| **Summer** | | | | | | | | | | |
| Summer 1 | M5 (NGC 5904) | Ser | GC | 15 | 18.6 | +02 | 05 | 5.8 | 17' | |
| Summer 2 | M13 (NGC 6205) | Her | GC | 16 | 41.7 | +36 | 28 | 5.9 | 16' | Great Hercules Cluster |
| Summer 2 | Rasalgethi | Her | ** | 17 | 14.6 | +14 | 23 | 3.5, 5.4 | 4.7" | |
| Summer 2 | M92 (NGC 6341) | Her | GC | 17 | 17.1 | +43 | 08 | 6.5 | 11' | |
| Summer 3 | M107 (NGC 6171) | Oph | GC | 16 | 32.5 | −13 | 03 | 8.1 | 10' | |
| Summer 3 | M12 (NGC 6218) | Oph | GC | 16 | 47.2 | −01 | 57 | 6.6 | 15' | |
| Summer 3 | M10 (NGC 6254) | Oph | GC | 16 | 57.1 | −04 | 06 | 6.6 | 15' | |
| Summer 3 | M9 (NGC 6333) | Oph | GC | 17 | 19.2 | −18 | 31 | 7.9 | 9' | |
| Summer 3 | M14 (NGC 6402) | Oph | GC | 17 | 37.6 | −03 | 15 | 7.6 | 12' | |
| Summer 4 | M80 (NGC 6093) | Sco | GC | 16 | 17.0 | −22 | 59 | 7.2 | 9' | |
| Summer 4 | M4 (NGC 6121) | Sco | GC | 16 | 23.6 | −26 | 32 | 6.0 | 26' | |
| Summer 4 | NGC 6231 | Sco | OC | 16 | 54.1 | −41 | 49 | 2.6 | 15' | |
| Summer 4 | M62 (NGC 6266) | Oph | GC | 17 | 01.2 | −30 | 07 | 6.6 | 14' | |
| Summer 4 | M19 (NGC 6273) | Oph | GC | 17 | 02.6 | −26 | 16 | 7.1 | 14' | |
| Summer 4 | M6 (NGC 6405) | Sco | OC | 17 | 40.1 | −32 | 13 | 4.2 | 15' | Butterfly Cluster |
| Summer 4 | M7 (NGC 6475) | Sco | OC | 17 | 53.9 | −34 | 49 | 3.3 | 80' | Ptolemy's Cluster |
| Summer 5 | M23 (NGC 6494) | Sgr | OC | 17 | 56.8 | −19 | 01 | 5.5 | 27' | |
| Summer 5 | M20 (NGC 6514) | Sgr | BN | 18 | 02.6 | −23 | 02 | 8.5 | 29' × 27' | Trifid Nebula |
| Summer 5 | M8 (NGC 6523) | Sgr | BN | 18 | 03.8 | −24 | 23 | 5.8 | 90' × 40' | Lagoon Nebula |
| Summer 5 | M21 (NGC 6531) | Sgr | OC | 18 | 04.6 | −22 | 30 | 5.9 | 13' | |
| Summer 5 | M24 | Sgr | OC | 18 | 16.9 | −18 | 29 | 4.5 | 90' | Small Sgr Star cloud |
| Summer 5 | M16 (NGC 6611) | Ser | BN+OC | 18 | 18.8 | −13 | 47 | 6.0 | 35' | Eagle Nebula |
| Summer 5 | M18 (NGC 6613) | Sgr | OC | 18 | 19.9 | −17 | 08 | 6.9 | 9' | |
| Summer 5 | M17 (NGC 6618) | Sgr | BN | 18 | 20.8 | −16 | 11 | 7.0 | 46' × 37' | Swan, Horseshoe, or Omega Nebula |
| Summer 5 | M28 (NGC 6626) | Sgr | GC | 18 | 24.5 | −24 | 52 | 6.9 | 11' | |
| Summer 5 | M69 (NGC 6637) | Sgr | GC | 18 | 31.4 | −32 | 21 | 7.7 | 7' | |
| Summer 5 | M25 (IC 4725) | Sgr | OC | 18 | 31.6 | −19 | 15 | 4.6 | 32' | |
| Summer 5 | M22 (NGC 6656) | Sgr | GC | 18 | 36.4 | −23 | 54 | 5.1 | 24' | |
| Summer 5 | M70 (NGC 6681) | Sgr | GC | 18 | 43.2 | −32 | 18 | 8.1 | 8' | |
| Summer 5 | M54 (NGC 6715) | Sgr | GC | 18 | 55.1 | −30 | 29 | 7.7 | 9' | |
| Summer 5 | M55 (NGC 6809) | Sgr | GC | 19 | 40.0 | −30 | 58 | 7.0 | 19' | |
| Summer 5 | M75 (NGC 6864) | Sgr | GC | 20 | 06.1 | −21 | 55 | 8.6 | 6' | |
| Summer 6 | NGC 6633 | Oph | OC | 18 | 27.7 | +06 | 34 | 4.6 | 27' | |
| Summer 6 | M26 (NGC 6694) | Sct | OC | 18 | 45.2 | −09 | 24 | 8.0 | 15' | |
| Summer 6 | M11 (NGC 6705) | Sct | OC | 18 | 51.1 | −06 | 16 | 5.8 | 14' | Wild Duck Cluster |

*(continued)*

| Sky Window | Object | Constellation[1] | Type[2] | Right Ascension hh | mm.m | Declination ° | ' | Magnitude | Size[3] | Comment |
|---|---|---|---|---|---|---|---|---|---|---|
| Summer 7 | Epsilon | Lyr | *** | 18 | 44.3 | +39 | 40 | 5.0, 6.1, 5.2, 5.5 | 208", 2.6', 2.3" | Double-Double |
| Summer 7 | M57 (NGC 6720) | Lyr | PN | 18 | 53.6 | +33 | 02 | 9.7 | 70" × 150" | Ring Nebula |
| Summer 7 | M56 (NGC 6779) | Lyr | GC | 19 | 16.6 | +30 | 11 | 8.2 | 7' | |
| Summer 7 | Collinder 399 | Vul | Asterism | 19 | 25.4 | +20 | 11 | 3.6 | 60' | Coathanger or Brocchi's Cluster |
| Summer 7 | Albireo | Cyg | ** | 19 | 30.7 | +27 | 58 | 3.2, 5.4 | 34" | |
| Summer 7 | M71 (NGC 6838) | Sge | GC | 19 | 53.8 | +18 | 47 | 8.3 | 7' | |
| Summer 7 | M27 (NGC 6853) | Vul | PN | 19 | 59.6 | +22 | 43 | 8.1 | 8' × 4' | Dumbbell Nebula |
| Summer 8 | M29 (NGC 6913) | Cyg | OC | 20 | 23.9 | +38 | 22 | 6.6 | 7' | |
| Summer 8 | M39 (NGC 7092) | Cyg | OC | 21 | 32.2 | +48 | 26 | 4.6 | 32' | |
| Summer 8 | NGC 6826 | Cyg | PN | 19 | 44.8 | +50 | 31 | 8.8 | 2' | Blinking Planetary |

**Autumn**

| Sky Window | Object | Constellation[1] | Type[2] | Right Ascension hh | mm.m | Declination ° | ' | Magnitude | Size[3] | Comment |
|---|---|---|---|---|---|---|---|---|---|---|
| Autumn 1 | Delta | Cep | Vr | 22 | 29.2 | +58 | 26 | 3.5-4.4 | 5.3 days | |
| Autumn 1 | M52 (NGC 7654) | Cas | OC | 23 | 24.2 | +61 | 35 | 6.9 | 13' | |
| Autumn 1 | NGC 7789 | Cas | OC | 23 | 57.0 | +56 | 44 | 6.7 | 16' | |
| Autumn 1 | NGC 457 | Cas | OC | 01 | 19.1 | +58 | 20 | 6.4 | 13' | Owl Cluster, Dragonfly Cluster, or ET Cluster |
| Autumn 1 | M103 (NGC 581) | Cas | OC | 01 | 33.2 | +60 | 42 | 7.4 | 6' | |
| Autumn 1 | NGC 869 | Per | OC | 02 | 19.0 | +57 | 09 | 5.3 | 30' | Double Cluster |
| Autumn 1 | NGC 884 | Per | OC | 02 | 22.4 | +57 | 07 | 6.1 | 30' | Double Cluster |
| Autumn 2 | M15 (NGC 7078) | Peg | GC | 21 | 30.0 | +12 | 10 | 6.4 | 12' | |
| Autumn 2 | NGC 7662 | And | PN | 23 | 25.9 | +42 | 33 | 8.9 | 32" × 28" | Blue Snowball |
| Autumn 3 | M73 (NGC 6994) | Aqr | Asterism | 20 | 58.9 | -12 | 38 | 9.0 | 3' | Only four stars |
| Autumn 3 | M72 (NGC 6981) | Aqr | GC | 20 | 53.5 | -12 | 32 | 9.4 | 6' | |
| Autumn 3 | NGC 7009 | Aqr | PN | 21 | 04.2 | -11 | 22 | 8.4 | 26" | Saturn Nebula |
| Autumn 3 | M2 (NGC 7089) | Aqr | GC | 21 | 33.5 | -00 | 49 | 6.5 | 13' | |
| Autumn 3 | M30 (NGC 7099) | Cap | GC | 21 | 40.4 | -23 | 11 | 7.5 | 11' | |
| Autumn 4 | M110 (NGC 205) | And | Gx | 00 | 40.4 | +41 | 41 | 8.0 | 10' × 5' | M31 companion |
| Autumn 4 | M32 (NGC 221) | And | Gx | 00 | 42.7 | +40 | 52 | 8.2 | 3' × 2' | M31 companion |
| Autumn 4 | M31 (NGC 224) | And | Gx | 00 | 42.7 | +41 | 16 | 3.5 | 160' × 40' | Andromeda Galaxy |
| Autumn 4 | M33 (NGC 598) | Tri | Gx | 01 | 33.9 | +30 | 39 | 6.3 | 60' × 35' | |
| Autumn 4 | M76 (NGC 650-1) | Per | PN | 01 | 42.2 | +51 | 34 | 11.4 | 3' × 1' | Little Dumbbell |
| Autumn 4 | Almach | And | ** | 02 | 03.9 | +42 | 20 | 2.1, 5.1 | 10" | Orange/blue |
| Autumn 4 | M34 (NGC 1039) | Per | OC | 02 | 42.0 | +42 | 47 | 5.5 | 35' | |
| Autumn 4 | Algol | Per | Vr | 03 | 08.2 | +40 | 57 | 2.1-3.4 | 2.867 days | Eclipsing binary |
| Autumn 4 | Alpha Per Cluster | Per | OC | 03 | 22.0 | +49 | | 1.2 | 185' | |

| Season | Name | Con. | Type | RA h | RA m | Dec ° | Dec ′ | Magnitude | Size | Notes |
|---|---|---|---|---|---|---|---|---|---|---|
| Autumn 5 | Mesarthim | Ari | ** | 01 | 53.5 | +19 | 18 | 4.6, 4.7 | 8″ | Orange/green |
| Autumn 5 | M74 (NGC 628) | Psc | Gx | 01 | 36.7 | +15 | 47 | 9.4 | 10′ × 10′ | |
| Autumn 5 | Mira | Cet | Vr | 02 | 19.3 | −02 | 59 | 2.0–10.1 | 331.96 days | Long-period |
| Autumn 5 | M77 (NGC 1068) | Cet | Gx | 02 | 42.7 | −00 | 01 | 8.9 | 6′ × 5′ | |
| **Winter** | | | | | | | | | | |
| Winter 1 | M45 | Tau | OC | 03 | 47.0 | +24 | 07 | 1.2 | 110′ | Pleiades, Seven Sisters, or Subaru |
| Winter 1 | Hyades | Tau | OC | 04 | 27.0 | +16 | | | | |
| Winter 1 | M1 (NGC 1952) | Tau | BN | 05 | 34.5 | +22 | 01 | 8.2 | 6′ × 4′ | Crab Nebula |
| Winter 2 | M42 (NGC 1976) | Ori | BN | 05 | 35.4 | −05 | 27 | 2.9 | 66′ × 60′ | Orion Nebula |
| Winter 2 | M43 (NGC 1982) | Ori | BN | 05 | 35.6 | −05 | 16 | 6.9 | 20′ × 15′ | |
| Winter 2 | Sigma | Ori | *** | 05 | 38.7 | −02 | 36 | 4.0, 7.5, 6.5 | 12.9″, 43″ | Triple star |
| Winter 2 | M78 (NGC 2068) | Ori | BN | 05 | 46.7 | +00 | 03 | 8.0 | 8′ × 6′ | |
| Winter 2 | Beta | Mon | *** | 06 | 28.8 | −07 | 02 | 4.6, 5.1 | 7″ | Triple star |
| Winter 2 | NGC 2244 | Mon | OC | 06 | 32.4 | +04 | 52 | 4.8 | 24′ | Rosette Nebula Cluster |
| Winter 2 | NGC 2264 | Mon | OC | 06 | 41.1 | +09 | 53 | 3.9 | 20′ | Christmas Tree Cluster |
| Winter 3 | NGC 1907 | Aur | OC | 05 | 28.0 | +35 | 19 | 8.2 | 7′ | |
| Winter 3 | M38 (NGC 1912) | Aur | OC | 05 | 28.7 | +35 | 50 | 6.4 | 21′ | |
| Winter 3 | M36 (NGC 1960) | Aur | OC | 05 | 36.1 | +34 | 08 | 6.0 | 12′ | |
| Winter 3 | M37 (NGC 2099) | Aur | OC | 05 | 52.4 | +32 | 33 | 5.6 | 24′ | |
| Winter 4 | M35 (NGC 2168) | Gem | OC | 06 | 08.9 | +24 | 20 | 5.3 | 28′ | |
| Winter 4 | NGC 2392 | Gem | PN | 07 | 29.2 | +20 | 55 | 8.3 | 13″ | Eskimo Nebula or Clown-Face Nebula |
| Winter 4 | Castor | Gem | ** | 07 | 34.6 | +31 | 53 | 1.9, 2.9 | 2.2″ | |
| Winter 5 | M79 (NGC 1904) | Lep | GC | 05 | 24.5 | −24 | 33 | 8.4 | 3′ | |
| Winter 5 | M41 (NGC 2287) | CMa | OC | 06 | 46.0 | −20 | 44 | 4.6 | 38′ | |
| Winter 5 | M50 (NGC 2323) | Mon | OC | 07 | 03.2 | −08 | 20 | 5.9 | 16′ | |
| Winter 5 | M47 (NGC 2422) | Pup | OC | 07 | 36.6 | −14 | 30 | 4.5 | 30′ | |
| Winter 5 | M46 (NGC 2437) | Pup | OC | 07 | 41.8 | −14 | 49 | 6.1 | 27′ | |
| Winter 5 | M93 (NGC 2447) | Pup | OC | 07 | 44.6 | −23 | 52 | 6.2 | 22′ | |
| Winter 6 | M48 (NGC 2548) | Hya | OC | 08 | 13.8 | −05 | 48 | 5.8 | 55′ | |

1. Constellation, see Appendix A.
2. Object type: **, double star; ***, multiple star; Vr, variable star; OC, open cluster; GC, globular cluster; BN, bright nebula; DN, dark nebula; PN, planetary nebula; Gx, galaxy
3. Apparent size of object in either minutes of arc, seconds of arc, or degrees. Most measurements were made from photographs; visual appearance may be smaller. For double and multiple stars, this number is a measure of the stars' separation from one another, while for variable stars, it lists the star's period of variability in days.
4. Most agree that M102 was, in all likelihood, a mistaken duplicate observation of M101. A few, however, have suggested that it might well be the galaxy NGC 5866 in Draco. Since NGC 5866 is a pleasant galaxy in its own right, I have chosen to include it here.

# Bibliography and References

## Books

Burnham, R., Jr. *Burnham's Celestial Handbook*, vols. 1, 2, and 3. New York: Dover, 1978.

Cherrington, E. *Exploring the Moon Through Binoculars and Small Telescopes*. New York: Dover, 1984.

Cosmolmagno, G., and Davis, D. *Turn Left at Orion*. New York: Cambridge University Press, 2000.

Dickinson, T. *Nightwatch*. Buffalo, N.Y.: Firefly, 1998.

Dickinson, T., and Dyer, A. *Backyard Astronomer's Guide*. Buffalo, N.Y.: Firefly, 2002.

Harrington, P. *Touring the Universe through Binoculars*. New York: John Wiley & Sons, 1990.

_____ , *Eclipse!*. New York: John Wiley & Sons, 1997.

_____ , *Star Ware*, 3rd ed. New York: John Wiley & Sons, 2002.

Houston, W., and O'Meara, S. *Deep-Sky Wonders*. Cambridge, Mass.: Sky Publishing, 1998.

Kepple, G., and Sanner, G. *Night Sky Observer's Guide*, vols. 1 and 2. Richmond, Va.: Willmann-Bell, 1998.

Mayall, N., et al. *Sky Observer's Guide*. New York: Western Publishing, 2000.

Pasachoff, J., and Menzel, D. *Peterson's Field Guide to the Stars and Planets*. Boston, Mass.: Houghton Mifflin, 1998.

Pennington, H. *Year-Round Messier Marathon Field Guide*. Richmond, Va.: Willmann-Bell, 1998.

Ridpath, I., et al. *Norton's Star Atlas and Reference Handbook*. Essex, U.K.: Addison Wesley Longman, 1998.

Tirion, W. *Sky Atlas 2000.0*, 2nd ed. New York: Cambridge University Press, 1998.

## Magazines

*Astronomy*, P.O. Box 1612, Waukesha, WI 53187; www.astronomy.com; monthly.

*Astronomy and Space*, Astronomy Ireland, P.O. Box 2888; Dublin 1, Ireland; www.astronomy.ie; monthly.

*Astronomy Now*, 193 Uxbridge Road, London W12 9RA, United Kingdom; www.astronomynow.com; monthly.

*Griffith Observer*, Griffith Observatory, 2800 East Observatory Road, Los Angeles, CA 90027; www.griffithobs.org; bimonthly.

*Mercury*, Astronomical Society of the Pacific, 390 Ashton Avenue, San Francisco, CA 94112; www.astrosociety.org; bimonthly.

*Sky & Telescope*, P.O. Box 9111, Belmont, MA 02178; www.skypub.com; monthly.

*Sky Calendar*, Abrams Planetarium, Michigan State University, East Lansing, MI 48824; www.pa.msu.edu/abrams; monthly.

*Sky News*, Box 10, Yarker, ON Canada K0K 3N0; www.skynews.ca; bimonthly.

*Stardate*, University of Texas, McDonald Observatory, Austin, TX 78712; www.stardate.utexas.edu; bimonthly.

## Software

*Deep Space* (CD-ROM; DOS), David Chandler Company, P.O. Box 999, Springville, CA 93265; www.davidchandler.com.

*Desktop Universe* (CD-ROM; Windows), Main Sequence Software, 102-100 Craig Henry Drive, Ottawa, ON Canada K2G 5W3; www.desktopuniverse.com.

*Deepsky 2000* (CD-ROM; Windows), Steven S. Tuma, 1425 Greenwich Lane, Janesville, WI 53545; www.deepsky2000.net.

*Earth-Centered Universe* (CD-ROM; Windows), Nova Astronomics, P.O. Box 31013, Halifax, NS Canada B3K 5T9; www.nova-astro.com.

*Guide* (CD-ROM; Windows), Project Pluto, 168 Ridge Road, Bowdoinham, ME 04008; www .projectpluto.com.

*Megastar* (CD-ROM; Windows), Willmann-Bell, P.O. Box 35025, Richmond, VA 23235; www .willbell.com.

*SkyMap Pro* (CD-ROM, Windows), World Wide Software Publishing, P.O. Box 326, Elk River, MN 55330; www.skymap.com.

*Starry Night* (CD-ROM; Windows or Macintosh), Space.com, 284 Richmond Street East, Toronto, ON, Canada M5A 1P4; www.starrynight.com.

*The Sky* (CD-ROM; Windows or Macintosh), Software Bisque, 912 Twelfth Street, Golden, CO; www.bisque.com.

*Touring the Universe through Binoculars Atlas* (CD-ROM; Windows), Phil Harrington and Dean Williams; www.philharrington.net.

*Voyager II* (CD-ROM; Windows or Macintosh), Carina Software, 12919 Alcosta Boulevard, Suite 7, San Ramon, CA 94583; www.carinasoft.com.

## Organizations

American Association of Variable Star Observers—25 Birch Street, Cambridge, MA 02138; www .aavso.org.

Astronomical League—11305 King Street, Overland Park, KS 66210; www.astroleague.org.

Astronomical Society of the Pacific—390 Ashton Avenue, San Francisco, CA 94112; www .astrosociety.org.

Royal Astronomical Society of Canada—136 Dupont Street, Toronto, ON Canada M5R 1V2; www .rasc.ca.

## Web Sites

| | |
|---|---|
| Abrams Planetarium | www.pa.msu.edu/abrams |
| *Astronomy Daily* | www.astronomydaily.com |
| *Astronomy* magazine | www.astronomy.com |
| Eclipse Home Page | sunearth.gsfc.nasa.gov/eclipse/eclipse.html |
| Jet Propulsion Laboratory's Comet Home Page | encke.jpl.nasa.gov |
| Phil Harrington's (author) Web site | www.philharrington.net |
| *Sky & Telescope* magazine | www.skyandtelescope.com |

# Glossary

**annual motion** The orbital motion of Earth as it completes a circuit around the Sun. As Earth orbits the Sun, the Sun appears to move through different constellations along the ecliptic as Earth orbits the Sun. Annual motion also allows us to see different stars and constellations at night during different seasons of the year.

**aperture** The diameter of a telescope's objective lens or primary mirror, usually expressed in inches, centimeters, or millimeters. The aperture of a refractor is measured according to the diameter of the objective lens, while the aperture of a reflector and catadioptric instrument specifies the diameter of the primary mirror.

**apparent size** How large an object appears in the sky. Apparent size is traditionally measured in degrees or in fractions of a degree, such as arc-minutes and arc-seconds.

**asteroid** An object made of rock or metal orbiting the Sun that is smaller than one of the planets. Most asteroids are located between the orbits of Mars and Jupiter.

**averted vision** A technique used by astronomers to see fainter objects in the sky. The retina of the human eye is made up of two types of light-detection cells, called cones and rods. Cones, which are concentrated toward the fovea (center of the retina), are sensitive to color and detail, but not to dim light. Rods, found around the edge of the retina, are the eyes' dim-light detectors. Looking a little to one side or the other of where a faint target lies, rather than staring straight at it, directs its weak light onto the peripheral area of the retina, where the light-sensitive rods stand the best chance of detecting it.

**big bang** A theory that explains the origin of the universe. It is believed that the universe was born some 15 to 18 billion years ago in a primeval explosion, which continues to expand to this day (apparently, accelerating in speed). Matter fragmented into huge masses, which eventually evolved into galaxies and in turn individual stars, planets, and you and me.

**bright nebula** A cloud of interstellar gas and dust that is visible through telescopes either because its hydrogen gas is excited into fluorescence by energy received from a nearby star or shines by reflected light from nearby stars. Nebulae that are excited into fluorescence are called emission nebulae or Hydrogen-II regions, while those that shine by reflected starlight are called reflection nebulae.

**catadioptric telescope** A hybrid design that combines the attributes of a refractor and a reflector. Light passes through a front lens, called a corrector plate, and on to a large concave mirror at the bottom of the tube, called the primary. Light then bounces off the mirror to a secondary mirror located just behind the corrector plate, then back through a hole in the primary mirror to the eyepiece. The most popular type of catadioptric among amateur astronomers is called the Schmidt-Cassegrain. Maksutov-Cassegrain telescopes are a second type of catadioptric telescope.

**celestial equator** The projection of Earth's equator against the backdrop of stars.

**celestial pole** The projection of Earth's rotational axis against the backdrop of stars. At present, Earth's axis, passing through the North Pole, is aimed very nearly at the North Star, Polaris; hence the North Celestial Pole is said to lie adjacent to Polaris. There is no "South Star," although the southern extension of Earth's axis points fairly close to a dim star in the constellation Octans.

**circumpolar** Those areas of the night sky near the celestial poles that neither rise nor set but are always visible above the horizon.

**comet** A small body composed of frozen gases and dust that revolves around the Sun, usually in a highly elliptical orbit. As a comet nears the Sun, solar radiation causes its frozen gases to sublimate, forming a large spherical cloud called the coma and frequently a tail.

**conjunction** A close apparent passage of the Moon or planet near another sky object as seen from Earth.

**constellation** A recognized pattern of stars in the sky. Moreover, astronomers have divided the sky into eighty-eight official constellations (listed in Appendix A). Most constellation patterns date back to the earliest civilizations. They represent something that our ancestors wanted to honor and remember for eternity, although few constellations actually resemble what or whom they memorialize.

**daily motion** The motion of the sky caused by Earth rotating once in twenty-four hours. Earth's daily motion causes the Sun, Moon, planets, and stars to appear to rise in the eastern portion of the sky and set toward the west.

**declination** One of two components of the celestial coordinate system, which specifies the exact position of an object in the sky. Declination is a measure of an object's position either north or south of the celestial equator. The celestial equator has declination value of 0°, while the North Celestial Pole has a value of +90° declination. The South Celestial Pole is found at declination −90°.

**deep-sky object** Broadly speaking, any object that lies beyond the solar system other than a single star. In the context of this book, the term "deep-sky object" applies to double and variable stars, star clusters, nebulae, and galaxies.

**degree** Distances between objects in the sky are measured in terms of angles. For instance, there are 90° from the horizon to the overhead point, or zenith. Each degree can be divided into sixty equal segments, called **arc-minutes.** The Moon, for instance, measures 0.5°, or 30 arc-seconds, across as seen from Earth. Each arc-minute can be further dividing into sixty equal segments, called **arc-seconds.** When astronomers want to specify the apparent size of a small sky object, they usually do so in arc-seconds.

**double star** Two stars that orbit around a shared gravitational center. Nearly half of the stars we see at night are double or multiple star systems, although most appear as single points of light to the eye alone. Binoculars or, more often, a telescope is needed to show the two (or more) individual stars in the system. The tighter the stellar pair, the higher the magnification and larger the aperture needed for resolution. No two double stars appear exactly the same. Many display striking contrasts in magnitude, while others are nearly equal in brightness. Some seem to shine pure white, while others glimmer with distinctive colors.

**eclipse** The partial or total obscuring of one celestial body by another, relative to an observer. A solar eclipse occurs when the New Moon passes between Earth and the Sun, while a lunar eclipse occurs when Earth's shadow falls across the surface of the Full Moon.

**ecliptic** The apparent annual path of the Sun through the sky. There are thirteen constellations along the ecliptic: Aries, Taurus, Gemini, Cancer, Leo, Virgo, Libra, Scorpius, Ophiuchus, Sagittarius, Capricornus, Aquarius, and Pisces.

**emission nebula** A cloud of interstellar gas that glows under its own power due to the ultraviolet energy imparted from a young star that lies within or nearby. The most famous example of an emission nebula is M42, the Orion Nebula, in the winter sky.

**eyepiece** A small lens or set of lenses that are inserted into a telescope and used to magnify and focus an image.

**field of view** The measure of how wide the view is through an eyepiece. When you look through an eyepiece, you see the scene that's in focus within a circular area surrounded by darkness. The angular width of that circular area, measured in degrees, is called the *apparent field of view*. The wider an eyepiece's apparent field of view, the more panoramic the effect. In the world of eyepieces, the apparent field of view typically ranges from a cramped 25° to a cavernous 80° or so.

Just how much of the sky will fit into an eyepiece's view can be estimated by dividing the eyepiece's apparent field (typically specified by the manufacturer) by its magnification through the telescope. This is known as the *real field of view*. For instance, let's imagine an 8-inch f/10 telescope and a 25-mm eyepiece. This combination produces 80×. Suppose this particular eyepiece is advertised as having a 45° apparent field of view. Dividing 45/80 shows that this eyepiece produces a real field of approximately 0.56°, a little larger than the Full Moon.

**finderscope**  A small, low-power, wide-field refracting spotting scope mounted piggyback on a telescope. Finderscopes help the observer aim the main telescope toward a target. They are usually specified by their magnification and aperture, such as 6 × 30 and 8 × 50, and have real fields of view between about 5° and 8°. Crosshairs in the eyepiece help an observer center the finderscope on a selected object. Before it can be used, however, the finder must be parallel with the telescope.

**focal length**  The distance from the objective lens or primary mirror to a telescope's focal point, the point where the light rays converge. In a reflector and a catadioptric, this distance depends on the curvature of the telescope's mirrors, with a deeper curve resulting in a shorter focal length. The focal length of a refractor is dictated by the curves of the objective lens as well as by the type of glass used to manufacture the lens. As with aperture, focal length is commonly expressed in inches, centimeters, or millimeters.

**focal ratio**  The focal length of a telescope divided by its aperture, often called a telescope's "f-number." An 8-inch telescope with a focal length of 56 inches would have a focal ratio of 7 (written "f/7"), since 56/8 = 7. Turning the expression around, a 6-inch f/8 telescope has a focal length of 48 inches, since 6 × 8 = 48.

**galaxy**  A huge system usually containing anywhere from millions to hundreds of billions of stars. Galaxies are further divided into three major classes, based on their overall structure and appearance. Spiral galaxies, like our own Milky Way galaxy, are huge pinwheel-shaped conglomerations of stars as well as star clusters and nebulae. Elliptical galaxies are spherical in appearance, curiously absent of nebulae but rich with globular star clusters. Finally, irregular galaxies fit in neither of the other categories and are often associated with powerful emissions of radio noise.

The universe is made up of billions of galaxies, with many forming in groups or clusters. The most famous galaxy cluster is found in the springtime constellations of Coma Berenices and Virgo; see chapter 6.

**globular star cluster**  A densely packed, globe-shaped collection of between 10,000 and 1 million stars gravitationally bound together and orbiting around the center of our Milky Way galaxy. Globulars are not unique to our galaxy, however. Large telescopes have detected them around many other galaxies as well. The stars in globular clusters are believed to be the oldest known, with some dating back 10 billion years.

**GoTo telescope**  A computerized telescope mount that will automatically steer a telescope toward a preselected target after the mounting has been initialized to time and location.

**Hydrogen-II (H-II) region**  An area of ionized hydrogen gas in space; see **emission nebula** for additional information.

**light pollution**  Light from streetlights, lights on buildings, and other earthly sources that shine skyward, where it is neither wanted nor needed, because of the offending fixture's poor design. Not only does this stray lighting wash out the stars, but it also adversely affects migrating birds and marine life, nocturnal animals, driver safety, public security, the environment, and in the case of municipal light, city budgets.

**light-year**  The distance that a beam of light will travel in the vacuum of space in one Earth year. Given that the speed of light equals approximately 186,000 miles per second (300,000 km per second), a light-year equals 5.87 trillion miles (9.45 trillion km).

**luminosity**  A measure of the inherent brightness of a star. The Sun appears very bright in our sky only because it is so close. Were we to pull away from the Sun a distance of a few light-years, it would quickly fade into the background because of its relatively modest luminosity.

**magnification**  The perceived increase in the size of a target when viewed through an optical instrument versus without. How much an image is magnified depends on the focal length of the telescope as well as the focal length of the eyepiece. If you look at the barrel of any eyepiece, you will notice a number followed by "mm." It might be "26 mm," "12 mm," or "7 mm," among others. This is the focal length of that particular eyepiece expressed in millimeters. Magnification is calculated by dividing the telescope's focal length by the eyepiece's focal length. Therefore, a 20-mm eyepiece in a 6-inch f/8 telescope (focal length = 48 inches) will magnify 61 times (written 61×). Remember to convert the two focal lengths into the same units of measure first (i.e., both in inches or both in millimeters). There are 25.4 millimeters in an inch.

**magnitude** A numerical measure of the relative brightness of an object in the sky, with brighter objects having lower numerical magnitude values. For instance, a 1st-magnitude star is about 2.5 times brighter than a 2nd-magnitude star, and about 6.3 times brighter than a 3rd-magnitude star. Very bright objects are assigned negative magnitude values, such as the Sun at magnitude −26 and the Full Moon at magnitude −13. Under very dark skies, the unaided eye can see stars as faint as magnitude 6 to 6.5, while 50-mm binoculars can show stars to about magnitude 9 on good nights.

**meteor** The rapid flash of a tiny grain of interplanetary dust burning up in Earth's atmosphere due to friction as it plummets downward. Meteors are sometimes called shooting stars or falling stars. While most completely burn up miles above Earth's surface, one will occasionally survive the incendiary plunge. If a meteor hits the ground, it is referred to as a meteorite.

**nebula** A cloud of interstellar gas and dust. Nebulae may be further categorized as bright nebulae and planetary nebulae.

**occultation** The passage of a large celestial object, most often the Moon, in front of a more distant object, such as a star or a planet.

**open star cluster** A loose collection of between a dozen and several hundred young stars. All of the stars in an open star cluster are believed to have formed from the same cloud of interstellar gas and dust. Open clusters are sometimes called galactic clusters, since they are located within the spiral arms of our Milky Way galaxy.

**opposition** The date when an object appears exactly twelve hours in right ascension away from the Sun. Dates of opposition are especially important when viewing the planets Mars and beyond, since those dates also signal when their distance from Earth is at a minimum and they, therefore, appear largest through telescopes. Amateur astronomers especially anticipate the dates when Mars is near opposition, since it shows the most surface detail at those times.

**planetary nebula** A shell of expanding gas exhausted from, and expanding around, an extremely hot star that is nearing the end of its life. Planetary nebulae often appear as small, greenish disks through small and medium aperture telescopes, which reminded early telescopic astronomers of the planets Uranus and Neptune and led to their misnomer.

**precession** A slow wobbling of Earth's rotational axis, taking 26,000 years to complete a gyration. As a result, the celestial poles move against the background stars. Although the North Celestial Pole is aimed toward Polaris at present, it hasn't always been that way. For instance, the North Celestial Pole was much closer to the star Thuban in Draco when the ancient Egyptians constructed the pyramids. Earth is not a perfect sphere but instead bulges by the equator, which causes this gyroscopic motion.

**reflection nebula** A dense cloud of interstellar dust that is illuminated by reflected starlight from an embedded or nearby star. Reflection nebulae tend to absorb the red end of the spectrum of light emitted from their illuminating stars, causing them to appear a deep blue in color photographs. One of the more prominent examples of a reflection nebula is M78 in Orion.

**reflector** A telescope that uses a large mirror at the bottom of the tube, called the primary, to gather light and bring it to a focus. Light bounces off that mirror to a secondary mirror and into an eyepiece for viewing. The most popular type of reflector among amateur astronomers is called the Newtonian reflector after Sir Isaac Newton, who is credited with its design.

**refractor** A telescope that uses a large lens in front, called the objective, to gather light and bring it to a focus, where an eyepiece is placed for viewing.

**resolution** The finest amount of detail visible in an image through a telescope or binoculars. For instance, resolution affects the visibility of the smallest crater seen on the Moon, the most subtle feature seen on a planet, or how close two stars can be to each other and still be seen as two separate points. Resolution improves as aperture increases, assuming the use of high-quality optics that are properly aligned.

**right ascension** One of two components of the celestial coordinate system, which specifies the exact position of an object in the sky. The sky is divided into twenty-four equal segments of right ascension, with each referred to as an hour. Beginning at the vernal equinox, the point in the sky occupied by the Sun on the first day of spring in the Northern Hemisphere, right ascension specifies an object's east-west position. The vernal equinox is located at 0 hours right ascension, with values increasing toward the east. The system wraps completely around the sky until the 24-hour right ascension line is reached, which coincides with 0 hours.

**seeing** A term used to rate the steadiness of star images seen through telescopes. A night is said to have "good seeing" if stars show little or no twinkling when viewed through a telescope. Those nights are ideal for planet watching, since turbulence in Earth's atmosphere can disrupt the view by making precise focusing difficult.

**star diagonal** A small prism or mirror assembly inserted into a refractor's or catadioptric telescope's focusing mount that is used to turn the eyepiece at a more comfortable angle for viewing. Most star diagonals turn the eyepiece at a 90° angle, in the process flipping the image left-to-right.

**star hopping** The technique promoted throughout this book for finding sky objects that are below naked-eye visibility. By aiming their telescope at a visible star, observers can take short hops from star to star while viewing through a finderscope, eventually winding their way to the desired target.

**supernova remnant** The expanding cloud of debris left over after a massive star has been consumed in a supernova explosion. After the explosion clears, all that remains of this once-massive star is the debris cloud and, in its center, the star's dense central core, called a neutron star.

**transit** The passage of a small celestial object in front of a larger object. Transits usually refer to Mercury or Venus crossing the face of the Sun, although 6-inch and larger telescopes can also show transits of Jupiter's moon across that planet's disk.

**transparency** A term used to rate the clarity of the sky. A night is said to be transparent if there is an absence of clouds, haze, fog, and mist. These often occur immediately after the passage of high-pressure weather systems, which are notable for their low humidity and dew points.

**unity finder** An aiming device mounted on the side of telescopes that projects a small red dot or a bull's-eye target onto a clear piece of glass, which acts as a beam splitter. The observer then sights through the glass, seeing both the reflected target rings as well as stars shining through from beyond. The first and still most popular unity finder sold is the Telrad.

**variable star** A star that changes in brightness. These fluctuations, which can take from minutes to years to occur, are caused for several different reasons. Some stars actually physically expand in size, rhythmically pulsing like a heart, or beating erratically and unpredictably. Other stars only appear to change in brightness, as an unseen companion star passes in front of the brighter member of a double-star system. Still other stars appear to dim as a passing cloud of carbon or other opaque material temporarily blocks their light.

**zenith** The point in the sky that is directly over an observer's head.

# Index

Italic page numbers indicate illustrations.

Printed in the USA
CPSIA information can be obtained
at www.ICGtesting.com
JSHW052005011124
72840JS00003B/26